CREATIONISM LAID BARE

CREATIONISM LAID BARE

10 CREATIONIST CLAIMS AND HOW TO REFUTE THEM

SERGE BOURGAINE

ORBIS
UNUM
PRESS

Library of Congress Control Number: 2020908324

ISBN 978-0-9994906-2-4 (ebook)
ISBN 978-0-9994906-5-5 (print book)

βopγηv03ρα

With heartfelt thanks to Michèle, Rod, and Mike
for their counsel, toil, and patience
on this and prior texts.

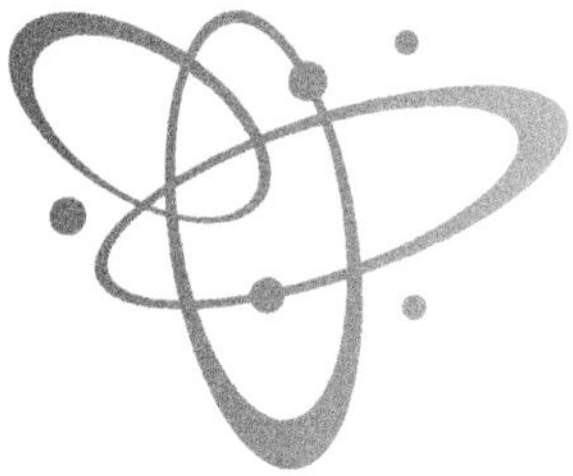

Contents

Prologue 1

I. The Point In Time Fallacy 13
II. The One Percent Fallacy 19
III. The Absolutes Fallacy 29
IV. The Circular Reasoning Fallacy 37
V. The Elevation Fallacy 41
VI. The Moral Supremacy Fallacy 47
VII. The Irrelevant Appeal Fallacy 55
VIII. The Post Hoc Fallacy 59
IX. The Wishful Thinking Fallacy 65
X. The Negative Proof Fallacy 77

Epilogue 83
Fallacies Redux 91
Fallacies Critiqued 95
A Letter To Creationists 107
A Letter To Charles 111
About The Author 119

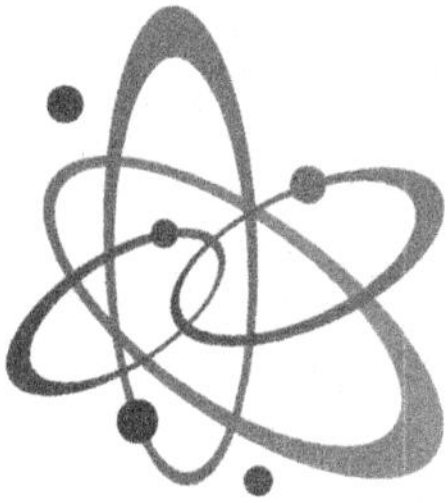

Prologue

The priests of the different religious sects dread the advance of science as witches do the approach of daylight.[1]
Thomas Jefferson (1743-1826)
American President

An unseen, supernatural entity forged planet Earth a mere six thousand years ago. The same mystical being created life in all its myriad forms, each attuned to its respective environment, each flawless in its design, each unchanging from generation to generation. Water stored in a chasm below the surface of the earth rose to inundate the planet in a Great Flood. The same water, as it receded into the abyss, carved the Grand Canyon. From the species who perished in the deluge, fossils of the best swimmers reside in the highest geological strata. Human footprints rest alongside fossilized dinosaur tracks. Radioactive decay sped up during the Biblical inundation, yielding the impression the earth is older than it really is. Continental drift occurred in months, not over millions of years. God fabricated light "in transit," thus all measurements to distant stars calculated from the speed of light and its redshift are incorrect. Isotropic properties of the universe attest to our planet's placement at the locus of the cosmos. The lifespan of Neanderthals was 300 years.

Creationist claims abound. The preceding sample illustrates their essence, their diversity, and their ingenuity. To the devout, they reflect divine truth - indisputable, eternal, and absolute. To the secular, they find expression from human minds both intelligent and fertile. To the skeptic, they expose two traits that underlie all creationist assertions. First, the absurdity of the pronouncements is surpassed only by their sophistry. And second, all such affirmations rely on fallacious reasoning. Yes, all. Every single one. It is this second trait that attracts our attention and inspires this text.

This is a handbook on reasoning. More precisely, a handbook on the exercise of fallacious reasoning as practiced by creationists. From this text, you will learn how to detect ten of the common arguments put forward by advocates of creationism. Equally, you will learn how to refute the arguments. This dual competency, practiced prudently and consistently, capably exposes the fragile foundation upon which all creationist doctrine rests.

The ten arguments that appear in this handbook emerged from a close examination of creationist assertions. Primary research spanned well over two hundred sources across four categories: reviews of scholarly papers written by creationists with scientific credentials; surveys of blog and website articles posted by creationist authors; synopses from dialogues with creationism's proponents; and parsings of public debates between, on the one hand, celebrated advocates of science and, on the other, staunch defenders of creationism. A tick mark on a log recorded each encounter with fallacious reasoning. A tally by type revealed the ten most frequent. Diagram 0.2, at the end of this *Prologue*, catalogs the ten.

Note I make no comment on whether creationism's widespread use of logical fallacies results from rectitude, artifice, or nescience. Such an inspection calls for closer scrutiny, the result of which bears on the competence and character of each claimant. It is a topic worthy of attention, but one best left in the hands of a skillful, non-partisan author. Here I attempt to focus on the fallacies, not their proponents.

Before we proceed to the fallacious arguments, it is necessary to complete several items of housekeeping. Quick flybys of creationism, logic, and the scientific method will set the stage for upcoming chapters.

CREATIONISM

Creationism is the Christian belief that the universe and all life arose from acts of divine creation. The originator, God, forged you as a favored species, monitors your existence, and periodically intervenes in world events. The Bible, as the true word of God, is the primary source of creationist beliefs. Codified dogma avers its historical accuracy.

I use *creationism* as a collective noun, encompassing various schools of creationist doctrine. Some schools, for example, assert a young earth; others an older but not too ancient a planet. Some embrace the Biblical flood; others temper its role. Some tolerate evolution; most abhor it. I also incorporate Intelligent Design and similar fundamentalist beliefs under the creationist umbrella, as their affinities outweigh their differences.

Not all Christians are creationists, but all major denominations of Christianity embrace at least some elements of creationist thought. As a general rule, the more conservative the denomination, the more likely it embraces creationism. Examples include the Seventh-Day Adventist Church, the Southern Baptist Convention, and Evangelicalism. Hinduism, Islam, Judaism, and other non-Christian faiths likewise espouse assorted dogmatic, creationist-like beliefs, each focused on their respective deities and doctrines. This text, however, focuses on Christianity.

SCIENTIFIC METHOD

Our second flyby outlines the scientific method, the primary mechanism by which science arrives at its conclusions. In the West, the method is a product of the Renaissance and the Age of Enlightenment, eras that marked the intellectual and cultural awakening that began in fourteenth-century Europe. Here is a 5-step summary of the method.

Ask The scientific method starts by posing questions about phenomena that one seeks to better understand. This first step, crucial for the advancement of knowledge, requires that inquiries be framed clearly and objectively. Sound inquiries employ unambiguous definitions, cite explicit premises, and permit alternative outcomes. Above all, however, competent inquiry demands open inquiry. Individuals must avoid queries explicitly designed to ratify existing beliefs. They likewise must be free to challenge deeply-held convictions. Societies or ideologies that fervently oppose diligent inquiry are unlikely to embrace the scientific method.

Predict The second step begins with the formulation of a hypothesis. A hypothesis is a suggested explanation derived from a small set of possibly correlated or causal data, each set pertaining to explicit scenarios. A hypothesis is a precursor to a theory, which follows later in the process. The principal attribute of a hypothesis is that it must be testable through observation and experiment. Making a hypothesis testable requires predictions. Consequential hypotheses predict novel phenomena; they do not merely elucidate existing phenomena.

Observe It is time to put the hypothesis to the test. This entails framing and executing experiments, robust in design and methodical in procedure, to ascertain the validity of the hypothesis. The nature of the experiments will vary by field of research, but all rely on the collection of empirical evidence. This step likewise includes analysis of the experimental results. Here, mathematics occupies a central role, isolating anomalies and connecting observations into intelligible patterns. The rigorous, skeptical, and dispassionate evaluation of evidence leads to a decisive juncture: the moment when the hypothesis is confirmed or rejected.

Verify This step demands the hypothesis be verified by someone other than its author - ideally, by independent actors conducting separate experiments in different locations. This reduces the risk that evidence results from chance, bias, error, inexperience, ineptitude, or deception. A broad set of confirmed hypotheses, once independently verified, become an accepted theory. In its scientific definition, a theory is a well-substantiated explanation of a wide range of phenomena. Each theory is founded on a comprehensive body of evidence, offering an assortment of testable and falsifiable predictions.

Accept At some point, a valid theory will have passed a range of testable predictions. It will have explained all the observations of reality that preceded it. It will have correctly predicted the results of novel phenomena. It will have passed the rigor of independent verification. The weight of evidence then compels acceptance of the theory,

> and it is deemed to be true. There is, however, a notable caveat. A theory might be rejected, despite a preponderance of substantiating evidence, if it runs afoul of tenaciously-held secular or religious beliefs. We touch on this caveat later in the text.

Two terms from the scientific method warrant reinforcement. Each is critical to our narrative. The first is *falsifiability*. Every hypothesis must be falsifiable. That is, there must be an experimental result that, if it occurs, disproves the hypothesis. It is impossible to overstate the importance of this requisite element. There is, in fact, no element of greater consequence in all of science. If a hypothesis cannot be invalidated, the hypothesis is unsound, the subsequent experiment will be unsound, and conclusions drawn from the experiment will be unsound. If I claim, for instance, that an invisible witch cast a spell on a redwood tree, causing it to mushroom to a remarkable height, the claim is not falsifiable. No test can refute the assertion. Hence the claim itself is not credible.

The second term is *theory*. In science, a theory is not a conjecture, hunch or whim. It is not built upon ingrained, unquestioned faith. It is not the product of deference to Holy Writ, prophetic assertions, or divine authority. It is not established, sustained, or vindicated "because a book says so." And it is never worshipped. A theory issues from the rigorous, transparent, and evidence-based methodology of science. It is the highest level of truth that science can offer, but it is not perfect truth. All scientific truths are provisional truths. Let me repeat this. *All scientific truths are provisional*. New evidence might challenge a theory - any theory - causing it to be revised or discarded. This is a regular and critical aspect of science. Science is a dynamic, self-correcting process, one that incessantly seeks new evidence to cultivate a superior understanding of phenomena.

Think about provisional truth another way. We all know water freezes at 0° Celsius. We also know that, under various circumstances, water remains liquid at even lower temperatures: in salty sea water, in flowing rivers, in clouds, and when supercooled in laboratories. From extensive experiment, scientists can predict fairly well the temperature at which water freezes in each circumstance. But this knowledge rests solely on experiments performed on and around a single, solitary orb: our planet Earth. From an understanding of atoms and their chemical bonds, we *think* that water behaves identically on all planets in the universe. But without visiting each and every one, we cannot *know* this with one-hundred percent confidence. As such, the freezing point of water is a provisional truth. It is founded on solid evidence, but one day could be revised or overturned. So again, *all scientific truths are provisional truths*.

LOGIC & LOGICAL FALLACIES

A fallacy, for our purpose, is simply an error in reasoning. Fallacious reasoning not only yields invalid inferences, it likewise blocks access to truth. The field that studies reasoning and its affiliated fallacies is logic, itself a branch of philosophy. Logic tends to classify reasoning into two broad categories, formal and informal. Here is some background.

Formal logic addresses the structure of an argument more so than its content. It also tends to focus on *a priori* arguments. Think of *a priori* reasoning as pure reasoning, independent from experiential or empirical knowledge. Mathematics is an exemplar of *a priori* reasoning. Mathematics can "reason" that 3 x (5 - 2) - 2 = 7, but it cannot, on its own, tell us the meaning of 7. It cannot assert there are seven planets in a solar system or seven legs on a horse.

Deduction, another technique often linked to *a priori* arguments, denotes a general-to-specific form of reasoning, where a set of general premises drives a specific conclusion. Deductive arguments tend to advance forward from cause to effect, where the validity of one or more premises drives the correctness of a conclusion.

Now, consider the second category of logic. Informal logic relies on *a posteriori*, inductive constructs. *A posteriori* reasoning emphasizes experience and empirical evidence, placing less confidence on pure rationalism. And induction, in direct contrast to deduction, is a specific-to-general form of reasoning. Inductive reasoning begins with specific observations, then formulates a general conclusion from the observations. In doing so, induction tends to move backward from effect to cause. *A posteriori* and inductive reasoning, as we will see, form a cornerstone of the scientific method.

Examples of deductive and inductive reasoning appear in Diagram 0.1. Aristotle (384-322 BCE), the father of deductive

Diagram 0.1 A synopsis of formal and informal logic, with examples of deductive and inductive reasoning.

	Formal	Informal
Principle	*A priori*, based on pure reasoning, independent from experience	*A posteriori*, based on experience and empirical evidence
Method	Deduction, moves from general to specific, from cause to effect	Induction, moves from specific to general, from effect to cause
	Example from Aristotle	*Example from Hume*
Premise 1	All men are mortal.	The Sun has risen in the east every morning.
Premise 2	Socrates is a man.	The Sun has not been seen to rise in the west.
Conclusion	Socrates is mortal.	The Sun will rise in the east tomorrow.

argument by syllogism, furnishes the first example.[2] David Hume (1711-1776), the British Empiricist philosopher, tenders the second.[3]

Logicians partition formal and informal logic into various subcategories. That said, there is no single, universal classification system - or even, as far as this author was able to determine - a set of common definitions. Sundry taxonomies exist, each relying on the expertise and predilection of its proponent. The number of discrete logical fallacies likewise varies by source, ranging from a handful to several dozen.

From the two main branches of logic, however, note this distinction. Formal logic, at least in principle, is bilaterally capable: it can either prove or disprove a claim. Disproof occurs when flaws are detected in the logical structure of an argument. The flaws, once disclosed, collapse the argument; they render it meaningless. Proof occurs when an argument's premises can be demonstrated to be irrefutably true. Valid premises, once confirmed, affirm a valid conclusion. In practice, though, authenticating the premises often is impossible.

Informal logic, by contrast, is only unilaterally capable. It can never prove a claim, not even in principle, as it never is possible to observe all pertinent evidence (here again, think about the freezing point of water). Induction can, however, handily disprove a claim. A single piece of contradictory evidence often suffices. This is the true power of informal logic, one that underlies the testability and falsifiability of all valid hypotheses.

Note that, in assessing an argument or a claim, it is possible for the two branches of logic to disagree. That is, an argument might pass a test of formal logic but fail an informal test, or vice versa. Consider this assertion: "If one does not accept the recent six-day creation understanding, then that person is actually not a Seventh-day Adventist."[4] It can be argued the assertion is logically valid, in that premise drives conclusion: If one rejects a core tenet of the church (the creation of the world in six, 24-hour days), then one cannot be a professing member of the church. Closer scrutiny, however, reveals informal fallacies. The most egregious one is that no evidence underlies the premise that creation took place in six days.

The preceding example highlights a consequential point, one that will be repeated throughout the text. In cases where the two branches of logic conflict, where formal and informal logic disagree on the validity of a claim, the final arbiter is evidence. Evidence is the gold standard. It is the *sine qua non* that distinguishes fact from fabrication. It is the adjudicator that bridges nations, cultures, and belief systems. And, notably, it is evidence that emerges from scientific discoveries that most often disproves creationist claims. Evidence, in short, has no equal

One final comment on logic. Its semantics, logical forms, and theories of inference are complex and well beyond the scope of this handbook. Yet they all serve a single goal: to construct a valid argument, one free from fallacies. It is that simple.

STRUCTURE

With our flybys now complete, let's examine the structure of this handbook. Ten chapters follow. Each presents one fallacy, subdivided into four parts. The first part defines both the fallacy and the creationist keywords that signal the fallacy. The second part provides an example. It presents a selected creationist claim that embodies the fallacy, followed by a Biblical verse characteristic of the type commonly touted to support the claim.

Part three illustrates the fallacy from beginning to end, along with its refutation. This section also depicts the workflow adopted by creationists to exploit the fallacy. The fourth and final part briefly highlights the scientific view on the topic at hand. Let's begin.

MISCELLANY

Notes that appear at the end of each chapter provide explanatory detail or occasional miscellany. Here is the first such note. I interchange the words *creationism*, *religion*, and *faith* throughout the text. This stems, in part, from the tendency of each religion or faith to incorporate at least some elements of creationist dogma. The interchange also curbs endless repetition of the word *creationism*.

Here are some additional creationist claims. The Second Law of Thermodynamics, which states that every system moves from order to disorder, disproves evolution. The probability that complex organisms could arise instantly by chance is nil. The fossils of marine invertebrates, those who lived nonmotile at the bottom of the ocean, will be found in the lowest geological strata. The fossils of birds, from their mobility, will be found in the highest strata. Nickel content in the sea, accumulated from river inflow, indicates the world's oceans are at most nine thousand years old. The entire earth's crust was formed from volcanic eruptions. Rates of continental erosion evince a youthful planet. The rate of decay of the earth's magnetic field attests the planet is no more than ten thousand years old. The present-day existence of gas-dissipating, short-period comets - which cannot survive more than ten thousand years - confirms a young earth. Extrapolating backwards from current global population levels to the first two humans, at a growth rate of one-half percent per year, attests that the earth is around four thousand years old. Mutations are always harmful; thus, random mutation cannot account for the origin of healthy species. If all peoples originated from a common ancestral population in Africa, all must have spoken the same language, but we see no evidence of this. Death, formerly an unknown phenomenon, entered into the world only with sin.

Common creationist fallacies and their triggers. Diagram 0.2

The Top Ten Creationist Fallacies

① Point In Time	Assumes what is not known today will remain unknown	*Science cannot explain ...*
② One Percent	Leverages scant findings to dismiss robust but unwelcome evidence	*Science cannot 100% prove that ...*
③ Absolutes	Asserts absolute truth and knowledge of that truth	*Mine is the one true faith*
④ Circular Reasoning	Restates a premise as a conclusion	*Scripture tells us that ...*
⑤ Elevation	Elevates faith, intuition, and emotion above reason	*Trust in faith*
⑥ Moral Supremacy	Defends a claim from a stance of moral superiority	*I am a Christian*
⑦ Irrelevant Appeal	Appeals to factors not relevant to a claim	*Jesus loves you*
⑧ Post Hoc	Asserts a claim from the order of events, not their causality	*Let us pray*
⑨ Wishful Thinking	Defends a claim from a strong desire for it to be true	*It must be true that ...*
⑩ Negative Proof	Declares a claim true because it has not been proved false	*Science cannot disprove that ...*

[1]

The Point In Time Fallacy

Voiced sardonically:
Let us work without reasoning; it is the only way to make life endurable.[5]
Voltaire (1694-1778)
French Philosopher

Definition

The proponent of the *point in time* fallacy presupposes that what is not known today will remain unknown in the future. Creationist assertions that employ the fallacy often begin with the words *science cannot explain ...*

Example

Creationists assert that *science cannot explain* the origin of the universe. This failure, they contend, opens the door to a faith-based explanation, permitting creationists to declare that the Biblical account of divine creation must be correct. Creationists affirm their account by way of Biblical verse: "In the beginning God created the heaven and the earth."[6]

Recognizing The Fallacy

POINT 1. Recognize that the creationist assertion, despite its verbiage, has nothing to do with the creation of the universe. Do not debate the origin of the cosmos. Here's why.

POINT 2. Recognize that science, for the lion's share of human history, was unable to credibly explain phenomena. It lacked the tools, the methodology, and the mindset. In the year 1200, for instance, science was incapable of explaining why leaves are green. The wave theory of light had yet to be discovered. Cell theory was unknown. And the decipherment of photosynthesis, the process whereby light energy is converted into chemical energy, lay centuries in the future.

POINT 3. Recognize that religion readily steps into scientific voids. The inability of thirteenth-century scientists to explain the color of leaves allowed creationism's forebears to intercede with their own arcane answer. Why are leaves green? It is the will of God

POINT 4. Recognize that a trite *Goddidit* offers no explanatory power. It sheds no light on the physical properties and molecular processes that cause leaves to be green. It merely expresses an incurious mindset, one trained to docilely accept theistic authority.

POINT 5. Recognize that as science advances, it becomes increasingly capable of explaining phenomena that once were mysteries. Yet, from its demand for empirical evidence and testable predictions, science progresses slowly. In this instance, seven centuries passed before science was able to explain why leaves are green. In due course, however, science gets there.

POINT 6. Recognize that when science arrives at a credible and robust explanation, the scientific theory invariably displaces religious dogma. History records countless examples. Here are three. Christianity asserted the sun orbits the earth; science confirmed the earth orbits the sun. Religion interpreted earthquakes as punishments; science demonstrated earthquakes arise from plate tectonics, where edges of the earth's plates slip past one another. Religion linked plagues to sin; science identified their pathologies and treatments.

POINT 7. Recognize that religion often resists new scientific explanations, seeking any conceivable loophole to sustain its mystical assertions. But slowly, inevitably, the scientific explanation prevails from weight of evidence.

POINT 8. Recognize that religion, as each of its chimerical claims is refuted, merely replays the fallacy. What will happen once science is able to explain the origin of the universe? The answer is simple. Creationists once again will scour the frontier of knowledge, singling out a different unexplained phenomenon. Perhaps they will select consciousness, as science cannot yet explain how consciousness arises from the trillions of neurons, synapses, and electrochemical signals that comprise each of our brains.

POINT 9. Recognize that creationists, from nothing more than cut-and-paste technology, then will fashion a new claim: *Science cannot explain ~~the origin of the universe~~ the rise of consciousness*. From this patched-up pretense, they will rerun the identical *point in time* fallacy. The sole purpose, once again, will be to assert the legitimacy of a favored Biblical tale.

POINT 10. Recognize that religion never concedes defeat. It never admits its charade. It never acknowledges its previous efforts to obscure the truth. So it proceeds, both *ad infinitum* and *ad absurdum*.

SUMMATION. Recognize that the original contention, that what is not known today will remain unknown in the future, is untrue. In actual fact, the very opposite is true. As science progresses, it becomes increasingly capable of explaining enigmatic phenomena. What is unknown today is likely to be known at a future date. Diagram 1.1 illustrates the creationist workflow for the *point in time* fallacy.

Science Responds

It is true that science cannot yet explain the origin of the universe. At present, science conjectures that the universe emerged some 13 billion years ago from a singularity, an infinitely dense and high-temperature point containing all the mass and energy of the universe, many times smaller than an atom. From causes not yet understood, the singularity began to expand. Within the first second of its expansion, the four fundamental forces of nature (the strong force, weak force, electromagnetic force, and gravity) emerged, and the first elementary particles (quarks) formed. Photons and atomic nuclei surfaced about ten seconds later and, along with other sub-atomic particles, dominated the universe. After 380,000 years, as the universe expanded and cooled, the first atoms (hydrogen) began to form. The same atoms, 300 million years later, coalesced into the first planets, stars, and galaxies. Nuclear processes within the stars forged heavier elements, such as carbon. When the most massive of the stellar objects reached the end of their lives, they exploded (as supernovae), spreading their matter into the expanse of space, to be incorporated into the next generations of stars and planets, such as Earth. And likewise into all lifeforms, including *Homo sapiens*. This is the Big Bang theory.

The Big Bang theory is a robust theory. It explains numerous facets of the universe, among them the prevalence of light elements, the large-scale structure of the cosmos, and the existence of cosmic background radiation. But it is not complete. Science cannot yet explain, for example, what led the Big Bang to bang. It likewise cannot elucidate every single property of the universe. The puzzling existence of dark matter and dark energy immediately come to mind. Despite the model's incompleteness, however, it possesses superior explanatory power and offers more testable predictions than any *ex cathedra* pronouncement of divine creation. And if the Big Bang theory one day is overturned, empirical evidence and falsifiable predictions - not otherworldly spirits - will inform its replacement.

MISCELLANY

Why are leaves green? The answer from physics, biology, and neuroscience. greatly simplified, is as follows. Plant and algae cells contain chloroplasts, specialized units that conduct photosynthesis. Chlorophyll, a pigment within the chloroplasts, absorbs red and blue wavelengths of light, but reflects green. Receptor cells in our eyes detect the reflected light, sending signals to our brains. Our brains, in turn, interpret the signals, creating our perception that leaves are green.

The creationist workflow for employing the *point in time* fallacy. **Diagram 1.1**

The Point In Time Fallacy

1. Select a phenomenon on the frontier of knowledge that science cannot yet explain.
2. Ascribe the phenomenon to the will of a deity, as a miraculous act beyond human comprehension.
3. Continue the charade until the day science develops a robust understanding of the phenomenon.
4. Resist the scientific explanation until a preponderance of evidence makes the religious claim untenable.
5. Go to step 1.

[II]

The One Percent Fallacy

If you wake up tomorrow morning thinking that saying a few Latin words over your pancakes is going to turn them into the body of Elvis Presley, you have lost your mind. But if you think more or less the same thing about a cracker and the body of Jesus, you are just a Catholic.[7]

Sam Harris (1967-)
American Neuroscientist

Definition

The advocate of this fallacy leverages anomalies and questionable evidence to assert a favored claim. He simultaneously discounts an extensive body of robust evidence that supports a rival but unwelcome claim. The fallacy takes its name from the relative proportions of desired but dubious evidence (1%) versus solid but unpalatable evidence (99%). The fallacy is triggered from the words *science cannot beyond all doubt prove that ...*

Example

Science estimates the age of the earth at 4.5 billion years.[8] Creationists assert that *science cannot beyond all doubt* prove the age of the planet, so they reject the claim. They allege a younger earth, one five to ten thousand years in age. Their arithmetic rests, most commonly, on the chain of genealogies from Adam to Jesus. The number and longevity of the individuals within the lineage inform the estimates. Creationists then adjust their numbers from various passages in Genesis, Exodus, and other books of the Bible. Some fine-tune their computations from the slavery of the Israelites,[9] from the construction of Solomon's temple,[10] or from the reign of a king. Here is one of many Biblical passages that aid the genealogical calculation: "And all the days that Adam lived were nine hundred and thirty years."[11]

Creationists, to buttress the Biblical claim, turn to pseudoscience. A favored gambit disputes the accuracy of radiometric dating, a technique that calculates the age of a substance by comparing the naturally-occurring elements in radioactive isotopes to their decay products. Creationists allege that radioactive decay "sped up" in the midst of the Biblical deluge, creating the perception the earth is older than it really is.[12]

Recognizing The Fallacy

POINT 1. Recognize that creationists seek anomalies or gaps in scientific understanding to craft rival claims. In this instance, creationists challenge aspects of radiometric dating - laboratory errors, experimental glitches, inconsistent results, varying decay rates of atoms - to denounce its precision.

POINT 2. Recognize that radiometric dating relies on properties within the nucleus of atoms and is not affected by outside temperatures or pressures, by magnetic or electrical fields, or by chemical or molecular properties. Recognize, too, that radiometric dating errors, where they exist, can be mitigated. Techniques include repeating a test, obtaining samples from diverse locations, and performing tests in multiple, autonomous laboratories.

POINT 3. Recognize that over forty types of radiometric dating methods exist. Each can corroborate another. Diagram 2.1 depicts examples of elements that can be used for dating, as well as their half-lives.[13] The age to which each element can date objects is multiple times its half-life.

Radioactive Isotopes

Diagram 2.1 The half-lives of various radioactive isotopes.

Isotope	Half-Life
Carbon-14	5,730 Years
Aluminum-26	740,000 Years
Iodine-129	17 Million Years
Samarium-147	108 Million Years
Uranium-235	704 Million Years
Potassium-40	1.3 Billion Years
Uranium-238	4.5 Billion Years
Thorium-232	14.0 Billion Years
Rhenium-187	41.6 Billion Years
Rubidium-87	49.0 Billion Years

POINT 4. Recognize that myriad non-radiometric dating methods also exist. Here is a sampling of their variety, along with the age (in years) to which they yield credible results: tree ring growth (11,000);[14] ice layering (150,000); luminescence dating (300,000); y-chromosome mutation (350,000);[15] erosion (6 million); seabed plankton layering (50 million); paleomagnetism (160 million); continental drift (200 million); impact crater frequency (300 million);[16] slowing of the earth's rotational speed (600 million); the recession rate of the moon (4 billion);[17] and gravitational lensing (14 billion).[18]

POINT 5. Recognize that all of the preceding techniques function as cross-checks. The age of an object as measured, say, by continental drift can be confirmed by radiometric dating, which in turn can be corroborated by paleomagnetism. No such empirical cross-checks exist among creationist claims.

POINT 6. Recognize that not all scientific dating methods can date an object back to billions of years. But all methods, encompassing still other measures of gradual natural processes - permafrost formation,[19] wood petrifaction, rock varnish accumulation,[20] and calcium carbonate stalactite growth[21] - disprove every young-earth hypothesis. Yes, every one.

POINT 7. Recognize, too, the evidence from extraterrestrial objects: asteroids, meteorites, and rocks returned from NASA moon missions. Plate tectonics - a process that on Earth reforms ancient rock - is absent from the objects. Thus, many off-earth bodies, unchanged since their formation, yield pristine samples. Dating this cosmic matter confirms it formed around the same time as the earth, 4.5 billion years ago.

POINT 8. Recognize that creationism regularly asserts pseudoscientific claims to proclaim both a young universe and an adolescent earth. Each ruse attempts to equate the interval of Biblical myth to the span of the earth's age. Creationists hope, by virtue of repetition, to instill confidence in the claims. Three such follow. Catastrophic plate tectonics induced continental drift to occur in months, not over millions of years, shaping the planet we see today.[22] Salt clocks, which measure the inflows, outflows, and current levels of sodium in the world's oceans, attest to a young earth.[23] A smaller than expected number of neutrinos emitted by nuclear fusion within the sun disproves an oldish star.[24]

POINT 9. Recognize that creationist assertions are assertions only, each averred by a shopworn *Goddidit*. All of which amounts to precisely nothing. There is no evidence that continental drift occurred in months, that oceanic salt clocks fix the age of the seas, or that neutrino emissions from the sun somehow prove a young earth. Each notional assertion merely attempts to eradicate truth at the whim of creationist dogma. Equally important, ample scientific evidence disproves each creationist claim.

POINT 10. Recognize that the book of Genesis, the impetus behind so many creationist claims, was written in an era of scientific ignorance. Consider further there is no sentence in the Bible that could not have been written by an author born in the prevailing age. This is odd. If Holy Writ reflects the infallible word of an omniscient deity, why is no omniscient knowledge explicitly disclosed? Yet none is. There is no pronouncement, no affirmation, no revelation that unambiguously reveals knowledge unavailable from that era. There is no equation, graph, or calculation that conveys a robust, mathematical comprehension of phenomena. No scripture discloses the testable tenets of an original, unheard-of, earth-shaking field of learning. Not one. Why not? The most plausible explanation is the self-evident one: Scripture and its purported truths arise in no respect from divine origins. They emerge merely from the fertile minds of *H. sapiens*. Scripture reflects nothing but the contemporary knowledge - and ignorance - of its sectarian authors.

SUMMATION. Recognize that the aim of the *one percent* fallacy is to discount large bodies of distasteful evidence. The guise permits creationists to exploit a smattering of anomalies, ones often invented by like-minded advocates, to cling to ingrained beliefs. Earth's age is but one example of this ubiquitous fallacy, where creationism attempts to dismiss extensive but unsavory evidence solely to maintain the fable of young planet. The creationist workflow for the *one percent* fallacy appears in Diagram 2.2.

Science Responds

No scientific law is correct 100% of the time. Take tides. Tidal behavior is well understood. Diurnal (one high and low tide per day), semi-diurnal (two high and low tides per day), and mixed (uneven) tides derive from the gravitational pull of the sun and the moon, and from the rotational properties of the earth and the moon.[25] Yet predictions of specific tidal action within the irregular landmass of a queerish cove might fail. There are always anomalies.

Creationism cleverly identifies anomalies, then sculpts its theories in alignment to its dogmatic ends. A gap in the fossil record, an inconsistency in radiometric dating, an unexplained fluctuation in gravity, a disagreement among evolutionary biologists - all serve creationism's dodge. The clever packaging of a cherry-picked set of such aberrations conveys a sense of credibility to the casual observer. Yet creationist scholarship is notable chiefly by its absence. Creationism generates little of its own evidence, preferring instead to critique the evidence of real science. Moreover, its theories invariably lack objectivity and rigor. Let's remind ourselves, first, that a theory is a well-substantiated explanation founded on an overarching body of evidence, offering testable and falsifiable predictions; and second, that a theory not only explains a broad range of phenomena but, equally, predicts the outcomes of new observations. Creationism's dirigiste assertions fall well below this bar.

Consider a rare instance of creationist research, one concerning helium in the earth's crust. From a single study, creationists cite the relative abundance of the element as evidence of a young planet. If the earth were billions of years old, they allege, all the helium would have dissipated via radioactive decay. Closer scrutiny, however, reveals sloppiness, both in experimental methodology and in interpretation of results. From the scant published data, mainstream scientists - physicists and geologists - refute the claim. Henke, for one, cites over eighty defects and open questions.[26] Here is a sample:

- Inconsistencies in the collection and storage of rock samples (e.g., maintaining samples at incorrect pressures and temperatures).
- Misidentification of rock samples.
- Misidentification of the purity of rock samples.
- Invalid diffusion measurements (measurements conducted in a vacuum, which draws gases out of rock at faster rates).
- Failure to distinguish between Helium-4 (caused by decay) and Helium-3 (not caused by decay) in the total helium measured.
- Failure to disclose possible contamination from outside helium (from thorium or uranium).
- Failure to disclose the potential impact of fluids altering rock cores.
- Unclear, incorrect, and incomplete statistical calculations (e.g., misuse of standard deviations and logarithmic values).
- No disclosure of raw data for evaluation by other scientists.
- No Independent verification of experimental results.
- No publication of experimental approach, analytical methodology, or results in mainstream, peer-reviewed journals.

The preceding affirms a pattern in creationist claims: limited original research, opaque methodology, exploitation of scientific anomalies, and fanciful interpretations - all orchestrated to preserve the beliefs of one particular faith.

Let's close this chapter with an acknowledgment. It is true that, on occasion, a new scientific theory arises from (or is confirmed by) an anomaly. One such incident involved gravity. Isaac Newton provided a comprehensive description of gravity in his masterwork, *Mathematical Principles of Natural Philosophy*, published in 1687. His depiction reigned supreme for over 200 years.

In the second half of the nineteenth century, however, it was observed that something was amiss with the orbit of the planet Mercury. Its movements did not quite follow the predictions of Newton's theory. The anomaly, among other factors, led Albert Einstein to produce a new theory of gravity, one that addresses conditions where gravity is intense and velocities approach the speed of light. His special and general theories of relativity not only explained Mercury's orbit, but introduced a new set of testable predictions.

Today, Einstein's theories are widely accepted. They extended Newton's work and, at the same time, forged a more robust model of the cosmos. Yet their acceptance stemmed from no alignment to a specific set of religious convictions. Their adoption, instead, derived from repeated experiments, each conducted time and again to an ever higher degree of accuracy by independent researchers, each conforming to a precise set of testable and falsifiable predictions. This is the high bar of evidence demanded by science.

MISCELLANY

A common creationist attack on radiometric dating, one not yet mentioned, asserts that no human from the ancient past witnessed the rate of radioactive decay. Therefore, no one can attest that radioactive isotopes decay at a constant rate. This "fact" allows creationists to claim that decay might have sped up during the Noahic Flood.

Creationism, in fact, extends this class of argument to nearly every unwelcome event from great antiquity: No human observed the event, ergo the scientific conclusion is untrustworthy. This argument is known as the *Were you there to see it?* fallacy. The artifice demands that a human observe an event, then credibly testify to the said event.[27] Creationists exploit the fallacy not only to attack unsavory scientific findings, but also to proclaim their own "evidence." They declare, for example, that contemporaneous humans attested to the historical accuracy of scripture.[28] They refer, of course, to the original authors of Biblical texts, to early apostles, and to Jesus himself - hardly impartial sources.

The *Were you there to see it?* argument is, to be sure, fallacious, as it suffers from multiple sophistries. Here I mention four. First, the demand for eyewitness testimony is merely a pretext, one that permits creationism to dismiss any objectionable fact from the distant past. Second, eyewitness testimony may be prominent within courts of law, but it is the lowest form of observational evidence in science.[29] Precision instruments and robust procedures - not the imprecision of human senses and human memory - drive the collection of sound observational data, as well as its analysis. Third, note that in *Were you there?*, the *there* is *here*. Past events, even ancient ones, leave traces - traces that can be experimentally examined in the current era. Think of fossil records, distant starlight, and the magnetic polarity embedded in igneous rocks (which, at the time the rocks formed, recorded the orientation of the earth's magnetic field). Think, too, of the relative abundance of non-radioactive to radioactive isotopes, and the decay products of the latter. And fourth, creationism offers no evidence whatsoever that radioactive decay sped up to just the right level - and for just the right duration - to confirm Biblical fictions.

The English polymath, Edmund Halley (1656-1742), appears to have been the first to suggest that the age of the earth's oceans could be estimated from their saltiness. The technique, though sound in principle, is unsound in practice. Sundry considerations annul salt-based measurements. To name a few. The original salinity of the sea, when the oceans first formed, is unknown. Salt clocks often assume that sodium, once deposited in the ocean, stays there (in fact, it is recycled). The rate of sodium deposit is not constant; it varies with rainfall, runoff, and erosion. Plate tectonics raise and lower seabeds, which in turn expose, melt, and subduct large concentrations of salt.[30]

The scientific evidence against creationism's delusory claims - accelerated continental drift, salt clocks, neutrino anomalies, and so on - is, in a word, overwhelming. An analysis of point by point refutations from mainstream science is well beyond the scope of this handbook. But the interested reader might wish to start with Dalrymple, Petto, or Hughes, whose respective texts provide excellent overviews.[31] Each furnishes sufficient detail both to disprove creationist assertions and to substantiate experimental results from legitimate science. Each also serves as a well-researched starting point for further reading.

From the following quote, it is tempting to think that Voltaire was commenting on the rigor of creationist research: "It is obvious the nose has been formed to bear spectacles - thus we have spectacles."[32]

The creationist workflow for employing the *one percent* fallacy. **Diagram 2.2**

The One Percent Fallacy

1. Scrutinize scientific findings in search of anomalies. If no anomalies emerge, consider inventing some.
2. Craft a desired claim from the selected anomalies.
3. Ignore, deny, or dismiss all evidence that refutes the claim.
4. Endlessly repeat the claim to instill its acceptance.
5. Preserve a stance of obscurantism, *ad infinitum*.
6. Go to step 3.

[III]

The Absolutes Fallacy

I would rather have questions I cannot answer
than answers I can't question.[33]
Max Tegmark (1967-)
Swedish-American Cosmologist

Definition

The proponent of this fallacy not only asserts the existence of absolute truth, but purports to possess knowledge of that truth. Words akin to *mine is the one true faith* invoke the fallacy.

Example

Creationists assert that their faith, and their faith alone, is true. From this premise, they portray the teachings of their faith as absolutes, each reflecting the eternal, unquestionable, and unwavering will of the Lord. Once again, creationists bolster their argument by quoting from scripture: "Thy word is truth."[34]

Recognizing The Fallacy

POINT 1. Recognize that if the creationist had been born in Sudan, he would be asserting the same truths about Muhammad or Allah that he does about Jesus. He would be citing, with equal zeal, not Biblical but Qur'anic verse.

Had he been reared in Punjab, he would be avowing - no less fervently - the dogma of Ik Onkar. Had he had been raised in ancient Rome, he would have reserved his highest reverence for the sky god Jupiter. If his family had lived among the pre-Columbian Maya, he would have shouted hosannas to Kinich Ahau. If his parents had resided in pre-Christian Scandinavia, he would have acclaimed the divine wisdom of Odin.[35]

POINT 2. Recognize, too, that from early history no deity spoke to all peoples worldwide. None revealed a consistent creation myth. None proffered an identical ethical code. None imparted the same assemblage of universal truths. None was everywhere affirmed as the one true god. Not one. Each of the countless chimerical deities, despite their putative omnipotent power, pandered only to the inhabitants of a narrow locality, to the same denizens who first forged each deity.

POINT 3. Recognize, from the preceding points, that faith derives in large part from the prevailing beliefs of a particular geography and era - the where and when of your birth. Throw in a few other variables - the beliefs of your parents, the beliefs of the local community, one's socioeconomic class, even one's gender - and your faith is all but determined. In other words, belief is relative. It is a function of time and place. It rarely derives from a neutral decision. It rarely is freely chosen.

POINT 4. Recognize that, as faith is relative, so, too, are the absolutes derived from each faith. They emanate not from divine will but from the creative minds who installed each faith. The absolutes persist, generation to generation, from each faith's contemporary interpreters - theologians, clerics, and evangelists.

POINT 5. Recognize that religion, ever eager to maintain its absolutes, relies on well-honed practices to indoctrinate its followers from formative age through adulthood. It inculcates though its clergy, its churches, its religious schools, its missionaries, its doctrines, its prayers, its catechisms, its charities, its symbolism, its moral codes, its rites of passage, its festivals, its feasts, its music, and its art. The targets of indoctrination, thus imbued to think in absolutes, become blind to the relative nature of their faith.

POINT 6. Recognize, as an epitome of a larger unity, the indoctrination of weekly church attendance. An authority figure calls the penitents to order. Congregants are instructed when to sit, when to stand, when to sing, when to be silent, when to open a sacred text, when to recite a holy passage, when to bow their heads, and when to pray. All present embrace the word of the Lord. There exists no diversity of thought, no questioning of authority, no deviation from the formula. Religion, admirably proficient in its indoctrination rituals, conducts them expertly and flawlessly, each burnished over millennia to a fine luster.

POINT 7. Recognize that religion, from its embrace of absolutes, conducts no vigorous scrutiny of its core beliefs. It resists deep introspection, meddlesome inquiry, and outside criticism, choosing instead to reside comfortably within its dogmatic, self-reinforcing edifice. Such flaws are endemic, as religion contains few of the checks and balances vital to self-reflection, self-correction, and critical thought. Religion admits no possibility its beliefs might be mistaken.

POINT 8. Recognize that religious absolutes, when testable, routinely are disproved or rebuked from want of evidence. Several examples evince the point: a deity stopping the sun;[36] stars orbiting in circles at uniform speed, not in ellipses at varying speeds;[37] a sea magically parted to allow crossing on dry land;[38] a river divided for an analogous purpose;[39] a horde of devil-infested swine committing suicide by drowning;[40] a man surviving the depths of the ocean in the stomach of a fish;[41] an ax with the ability to swim;[42] and the affirmation of a fiery purgatory awaiting the unrepentant sinner.[43]

POINT 9. Recognize that, when confronted with the fact that religious passages contradict all known physical reality, religion seeks an escape route. After centuries averring the literal and unassailable truth of said passages, persecuting anyone who dared challenge their veracity, religion adopts a new tack. It conveniently and unashamedly converts once-sacred but now-disproved precepts into handy parables. Each is deftly rehabilitated to reinforce moral or religious principles. Clever minds, as a case in point, often recast the deity-stops-Sun fable as a comforting allegory: *The Lord stands with you in time of need*. This act, at base, adroitly transforms testable predictions into non-testable stratagems - an action that, perhaps more than any other, belies the existence of religious absolutes.

SUMMATION. Recognize that faith is relative, tethered to time and place. Its absolutes, equally so. Indoctrination sustains the absolutes. From inculcation, each faith asserts that its absolutes - and its absolutes alone - are true, dismissing the verities of rival faiths. Yet religious absolutes, as unalterable to the believer as pliant to the skeptic, serve solely as wishful, evidence-free constructs to sustain particular beliefs. They in no way substantiate faith-based claims. Diagram 3.1 summarizes the creationist workflow for the *absolutes* fallacy.

Science Responds

As more than one prominent psychologist has noted, the attractiveness of religious belief derives, in large part, from its ability to satisfy our psychological needs.[44] Among them, a search for meaning, status, order, social acceptance, and emotional connection. Religion, well aware of this reality, administers a potent palliative - absolutes. Absolutes impart certainty. Absolutes render comfort. Absolutes provide structure. And to the devout, absolutes promise future reward. Note, however, the palliative obtains its effect from *belief* in the absolutes, not from their veracity.

Psychology further confirms that the human mind, ever susceptible to what it wants to believe, all too often embraces religious absolutes with little scrutiny. It then labels the absolutes as imperishable truths, ardently defending them to protect an improvised mental world.

Paleontology, sociology, neuroscience, and evolutionary biology - among other sciences - tender as explanation both competing and complementary conjectures. Some concern the survival instincts of our paleo-ancestors and the structure of their societies. Others derive from the physical and chemical make-up of our brains. All attest, in one form or another, to a proclivity of our species to enshrine preferred beliefs. And, once installed, to sustain them even in the face of contradictory evidence.

Yet here's the rub. Faith as a path to mental comfort is irrefutable. But faith as an indicator of external truth, temporal or otherwise, is fallacious. Faith-based thinking - founded on revelation, scripture, prayer, and intuition-as-truth - affords no predictive or expository power. None. In this context, religion's talent for affirming absolutes derives from the most rudimentary, quotidian explanation: Mythomanes simply fabricate the truths, each free from any burden of evidence, to satiate our mental anxieties.

MISCELLANY

Here, I would ask the apostle to dispassionately assess religion's track record in explaining the world. I request he evaluate its myriad predictions of the second coming of Christ, of the end of the world, and of other religiously significant events. What percentage of the predictions proved to be true? If his analysis is unbiased, he will arrive at a single, salient conclusion: Religion's history of successful predictions is, bluntly put, abysmal. Here is a further sample of testable, religious prophecies that, in their day, were held as inspired truths. Yet each has been unequivocally refuted. The speed of light is instantaneous.[45] All planets reside in seven spheres.[46] Angels propel an eighth sphere, to which the stars are attached.[47] A lighter body of water exists above the celestial firmament.[48] The earth rests at the center of the cosmos.[49] The first plants took root within seventy-two hours of the earth's existence.[50] The sun, a mere ninety-six hours after its formation, emitted its first light.[51] The firmament is eternal and unchanging.[52] Woman was made to be a helper to man.[53] All material creatures are ruled by inferior angels who, in turn, are directed by superior angels.[54]

Here are additional creationist assertions that gainsay all physical reality. God created both light[55] and plants[56] before the sun. God created both earth[57] and fruit trees[58] before the stars. God fashioned plants before the rain,[59] earth before the moon,[60] and oceans before the land.[61] God hatched birds before insects[62] and formed whales before reptiles.[63] He also forged a non-moving earth.[64]

And here are still more discredited predictions. Protestantism will collapse by 1825.[65] Jesus will return in his lifetime.[66] Jesus will return in 70 CE.[67] Jesus will return in 400.[68] Jesus will return in 1186.[69] Jesus will return in 1284.[70] Jesus will return in 1533.[71] Jesus will return in 1843.[72] Jesus will return in 1889.[73] Jesus will return in 1935.[74] Jesus will return in 1936.[75] Jesus will return in 1981.[76] Jesus will return in 1988.[77] Jesus will return in 2007.[78] Jesus will return in 2010.[79] And so on.

The creationist workflow for employing the *absolutes* fallacy. Diagram 3.1

The Absolutes Fallacy

1. Declare and adopt a set of absolutes.
2. Assert claims from your absolutes. Pretend your absolutes, and your absolutes alone, are real.
3. Refuse to accept that, had you been born elsewhere, you would just as passionately assert a different set of absolutes.
4. Refuse to accept that both faith and absolutes are relative, linked to the era and geography of your birth.
5. Refuse to accept that individuals are indoctrinated both into their faith and into their absolutes.
6. Refuse to accept that, when testable, absolutes routinely are disproved.
7. Go to step 2.

[IV]

The Circular Reasoning Fallacy

The more the concept of reason becomes emasculated, the more easily it lends itself to ideological manipulation and to propagation of even the most blatant lies.[80]

Max Horkheimer (1895-1973)
German Philosopher

Definition

Circular reasoning declares a premise valid, then presents a conclusion as a restated form of the premise. The fallacy takes the form *A is true because A is true*. The sophistry often begins with the phrase *scripture tells us that ...*

Example

Creationists proclaim the Bible as the true word of God. From this presumption, creationists assert the truth of scripture. This translates into a self-verifying *scripture is true because scripture is true* argument. Creationists buttress their argument by quoting various passages from the Bible. Among them: "Every word of God is pure";[81] "And now, O Lord God, thou art that God, and thy words be true";[82] and "all scripture is given by inspiration of God."[83]

Recognizing The Fallacy

POINT 1. On no account debate scripture. It is a fool's errand, as context and interpretation vary widely. Moreover, scripture-based claims, both malleable and evidence-free, invariably conform to the whims of the advocate.

POINT 2. Recognize that circular arguments only pretend to establish a claim, as the conclusion simply is a restated form of the premise: *Scripture is true, so derive truth from scripture*.

POINT 3. Recognize that circular arguments often stress a key term to render a claim impervious to disproof. Note this declaration: A *true* believer accepts the truth of scripture. The key term *true* shifts any uncertainty regarding the validity of scripture to the ignorance or obstinance of the doubter.

POINT 4. Recognize that among all fallacies exploited by creationists, *circular reasoning* is by far the most common. It underlies countless dogmatic claims. Recognize, too, that circular reasoning invariably is invalid. Yes, invariably.

SUMMATION. Recognize that, in creationism, a circular argument is a bid to assert a self-verifying claim, one that disallows contradiction, renounces empiricism, and repudiates independent verification. As such, all circular arguments become meaningless tautologies. The creationist workflow for the *circular reasoning* fallacy appears below.

Science Responds

Science need not respond, as all circular reasoning is invalid

The creationist workflow for employing the *circular reasoning* fallacy. **Diagram 4.1**

The Circular Reasoning Fallacy	
①	Declare the existence of the one true God.
②	Declare scripture as the inerrant word of God.
③	Combine ① and ② into a self-verifying logical fallacy: Holy Writ is the word of God, hence scripture is infallible.
④	Emphatically proclaim the fallacy.
⑤	Go to step 4.

[V]

The Elevation Fallacy

Christian philosophy bids [reason] give place and yield complete submission to the Holy Spirit.[84]

John Calvin (1509-1564)

French Theologian

Definition

The proponent of this fallacy attempts to defend a cherished claim by elevating faith above reason. The advocate likewise cultivates various guises to dismiss, alter, or ridicule any and all evidence that might refute the claim. The phrase *trust in faith* unmasks the fallacy.

Example

The theologians Augustine of Hippo[85] (354-430) and Thomas Aquinas[86] (1225-1274), among others, set the tone of fideism by asserting that faith stands above reason. Pope Pius IX (1846-1878) echoed the stance in his *Syllabus of Errors*, published in 1864. The *Syllabus* condemned eighty errors of thought that contravened Church doctrine.[87] Error number eight is ranking reason equal to religion. To further underscore the authority of the Church, in 1870 the First Vatican Council declared the pope infallible. Anyone who challenged the word of the Roman Pontiff was, and is, deemed "anathema" - cursed.[88] More recently, in 2009 Pope Benedict XVI (1927-) re-affirmed that "Reason always stands in need of being purified by faith."[89]

Protestant theologians, not to be outdone by their Catholic brethren, assert claims of similar form. Note this exhortation from Martin Luther (1483-1546), an architect of the Protestant Reformation: "Reason is the greatest whore of the devil."[90] Again, creationists turn to scripture to affirm their stance: "For whatsoever is not of faith is sin."[91]

Recognizing The Fallacy

POINT 1. Recognize that church leaders routinely denounce reason. They rightly and astutely interpret reason as a threat to faith. To sustain creationism, reason must be constrained.

POINT 2. Recognize that faith-as-truth arises from revelation, scripture, prayer, and the whims of an invisible entity - methods orchestrated to deprecate reason and evidence.

POINT 3. Recognize that to assert its numinous truths, religion not only hefts faith above reason, but also elevates intuition, emotion, and "feel-goodism" as sources of spiritual and temporal truth: "Trust in the LORD with all thine heart." [92] Each such pronouncement serves to belittle or suppress reason.

POINT 4. Recognize that religion leverages the legions of inconsistencies within scripture to unearth verses that support its dogma, yet disregards or reinterprets unwelcome passages. Note this instance concerning faith: Some passages assert that salvation is justified by faith alone.[93] Yet others expressly state good works are required, that faith alone is insufficient.[94]

POINT 5. Recognize that religion cherry-picks scientific evidence in alignment with its dogma. As a case in point, creationists affirm the expansion of the universe, noting that the Lord "created the heavens, and stretched them out."[95] Then, from the very same expansion, creationists adopt a diametric stance. They spurn all scientific measurements of the universe's rate of expansion. The reason? All such measurements disprove a young earth.

POINT 6. Recognize that, where testable, faith-based assertions are routinely disproved. Consider again the fable of a deity stopping the sun to extend daylight.[96] Halting the sun - either its rotation about its axis or its revolution around the Milky Way galaxy - would have no effect on the measure of daylight falling on the earth. It is odd that an allegedly omniscient deity would base his dubious deed on the disproved theory of geocentrism.

POINT 7. Recognize that religion often disregards the unremitted refutation of its prophecies, branding each disproof as a test of faith. This technique permits the believer to brush aside any facts, people, or practices that raise uncomfortable truths. Faith, in this context, serves as nothing more than an enchanted token, shielding the intransigent disciple from any evidence that threatens a cherished belief.

POINT 8. Recognize that, more broadly, religion deliberately designs faith to be untestable, then proudly proclaims this same non-testability as a virtue of faith. The principle of revealed truth, for example, accords to the tractable advocate a get-out-of-jail-free card, one that permits its holder to herald any claim - wholly destitute of testable and falsifiable conditions - as the emblazoned and inerrant wisdom of a deity.

POINT 9. Recognize that, when pressed, religion invokes a form of the *God of the Gaps* argument, asserting that any gaps in knowledge unexplained by science *ipso facto* attest to the existence of a deity. This, in essence, divides knowledge into separate realms of faith and reason. By proclaiming the gaps as the realm of a deity, religion endeavors to maintain its authority on metaphysical and theological subjects. This argument, however, creates a difficulty for creationists. As science advances and explains ever more phenomena, the deity is squeezed into ever-narrowing gaps and, ultimately, into a final redoubt.

SUMMATION. Recognize that religion employs assorted stratagems to elevate faith above reason, the intent of which is to alter, dismiss, discredit, or cleanse all unwelcome evidence. For its own survival, it has no other choice. It knows that a battle waged on reason and evidence will be lost. The creationist workflow for the *elevation* fallacy appears in Diagram 5.1.

Science Responds

A teapot floats right behind and just above your head. It has always been there and always will be, throughout your entire life.[97] It is a beautifully painted, pale blue, ceramic teapot. It happens the teapot is endowed with a unique property. You cannot sense it in any way. If you turn your head, the teapot shifts its position, remaining just outside your field of vision. If you look in the mirror, the vessel reflects no light. If you reach behind your head and attempt to strike the teapot with a hammer, it vanishes, only to reappear when the danger has passed. No empirical test discloses its existence. Only your faith tells you it is there. No one else can sense the teapot, but you know it is there, floating at this very moment behind your head.

The teapot claim, as just articulated, yields two insights. First, it mimics the practice of religion. It elevates faith above reason to assert the existence of a dubious claim. It is faith alone, absent of all evidence, that sustains belief in the teapot's existence. Second, the claim is untestable. Purposefully so. The stated premise - that no empirical evidence can reveal the floating vessel - nullifies all objective tests.

Designing claims to be untestable is a universal stratagem of creationism, one that underlies countless faith-based assertions. Science recognizes, however, that the assertion of a floating teapot is as valid as any non-testable religious belief. And just as worthy of dismissal.

MISCELLANY

Irony [noun; origin from French *ironie*, Latin *ironia*, Greek *ειρωνεία* for "dissimulation, assumed ignorance"]: a humorous or sarcastic use of words, or a juxtaposition of ideas, that draws attention to some incongruity or irrationality.[98] Sample usage: Creationists, in their zeal to safeguard conjured truths, routinely extol the untestable tenets of their faith. Yet at the same time, though loathe to admit it, they mutely suffer from science-envy, begrudging both the transparency and nonpareil capability of the scientific method. The irony? As Sagan affirmed, "There isn't a religion on the planet that doesn't long for a comparable ability - precise, and repeatedly demonstrated before committed skeptics - to foretell future events."[99]

The creationist workflow for employing the *elevation* fallacy. **Diagram 5.1**

The Elevation Fallacy

1. Elevate faith above reason and science, branding the latter two as untrustworthy.
2. Equally elevate intuition, emotion, and "feel-goodism" above reason and science.
3. Leverage inconsistencies in scripture, affirming only those verses and interpretations that support the avowed dogma.
4. Label unwelcome evidence as a test of faith, and dismiss it.
5. Construct religious truths to be untestable, hence irrefutable.
6. Go to step 1.

[VI]

The Moral Supremacy Fallacy

Persecution is used in theology, not in arithmetic,
because in arithmetic there is knowledge,
but in theology there is only opinion.[100]
Bertrand Russell (1872-1970)
British Philosopher

Definition

The proponent of this fallacy asserts the truth of a claim from a stance of moral supremacy. The words *I am a Christian* fuel the fallacy.

Example

Creationists assert that God is the source of morality, leading to the tautology that morality is the application of God's laws as expressed in scripture (see the *circular reasoning* fallacy). It naturally follows, they contend, that belief in the one true God endows Christians with a stronger moral compass than non-believers.

They extend the corollary by averring that Christians, and Christian societies, are less prone to underwrite violence. As evidence, creationists cite the crimes of twentieth-century despots - Hitler, Stalin, Mao, Pol Pot - as the inevitable acts of non-Christian authority. Creationists often quote one Biblical passage above all others as the source of Christian morality: "Thou shalt love thy neighbor as thyself."[101]

Recognizing The Fallacy

POINT 1. Recognize that non-Christians corner no monopoly on violence. Christians, and Christian societies, possess their own extensive catalog of heinous acts.

POINT 2. Recognize that numerous Biblical passages explicitly advocate violence. Two evince the point. In the Old Testament, God declares, "Now therefore kill every male among the little ones, and kill every woman that hath known man by lying with him. But all the women children, that have not known a man by lying with him, keep alive for yourselves." [102] In the New Testament, consider the fiat from Jesus, directed at those who abstained from his authority: "But those mine enemies, which would not that I should reign over them, bring hither, and slay them before me."[103]

POINT 3. Recognize that, throughout history, believers have conducted interminable vile acts - Catholics butchering pagans, Protestants burning Catholics, Christians exterminating Jews - in country after country, in century after century, all expressly in the name of Christianity. Consider, too, that Christianity participated in the extermination or forced conversion of entire civilizations in Africa, Asia, Latin America, and North America.

POINT 4. Recognize that religion often accredits its most heinous acts, as well as the fealty of their messianic perpetrators, to the truth of their faith and the supremacy of their deity.

POINT 5. Recognize that creationism's literal interpretations of Biblical verse condemn billions of innocents to Hell. The reason? Each was born into a non-Christian society, with little or no opportunity to embrace Christian teachings.

Each failed to accept Jesus as Lord and Savior. Each thus consigned himself to eternal perdition, solely from a mere accident of birth. And according to creationists, of the nine million children who die worldwide every year before the age of five, nearly all will go to Hell. Again, the reason? Because most never were baptized in the name of the Father, the Son, and the Holy Ghost."[104] Are these the acts of a benevolent deity with the power to change this reality, but who elects not to?

POINT 6. Recognize that many Christian faiths embrace predestination, the belief that God predisposes the soul of each individual to be saved or to suffer eternal damnation. If one's destiny is predetermined at or before birth, the existence of arbitrary suffering under the guise of a beneficent deity is even less defendable.

POINT 7. Recognize that monotheism - the belief in one, and only one, deity - breeds divisiveness. It induces an absolutist, either-or mentality, one that advocates "my deity is valid and yours is false." The Crusades, for example, championed the "truth" of Christianity over the "heresy" of Islam. From comparative studies of historical conflicts, more than one scholar concludes that monotheistic gospels eclipse their polytheistic and nonreligious competitors in admonishments to destroy infidels, smite apostates, and cast unbelievers into the bowels of hell - relegating all such wrongdoers to an eternity in the infernos of the netherworld.

POINT 8. Recognize that theocratic societies, past and present, frequently operate oppressive, intolerant regimes. Compare them to secular societies, ones that uphold the separation of church and state. Where would you rather live? Consider further that countries whose denizens express low levels of religious belief - Denmark and Sweden, for instance - often score highest on international measures of social equality, societal stability, and happiness.[105]

POINT 9. Recognize that religionists routinely credit the beauty of creation to their god but, to the same deity, never apportion blame for the world's horrors. Instead, they ascribe famines, plagues, tsunamis, and other dire acts to "mysteries" of creation.

POINT 10. Recognize that all of the preceding points, while providing valid counterarguments to claims of a superior Christian morality, miss the salient point. The oft-neglected fact is this: Atrocious acts commonly stem not from religion but from absolutism. Whether the absolutist ideology is secular (we are the master race) or religious (we are the chosen people), the result is identical. Each oppressor translates nativist motives into an absolutist philosophy. One that avers incontrovertible truth. One that tolerates no deviation from coveted aims. One that brooks no defiance to the will of the leadership. One that obligates the fanatical destruction of its opponents. And one that stands all too ready to impose its tainted morality on everyone. It is this expression of absolutist ideologies - any absolutist ideology - that attracts like-minded followers and sows the seed for many of humanity's most horrendous acts. It is this consolidation of power into absolutist hands by intransigent, demagogic, or totalitarian leaders that permits the seed to germinate. It is, thus, all absolutist ideologies, including those from religion, that set the stage for potential atrocity.

SUMMATION. Creationism often fails to comprehend, from its assertion of an absolutist ideology, that it, too, cultivates violence. Christian leaders and laity alike have leveraged their inviolable tenets to ridicule, censure, ostracize, threaten, mistreat, torture - even murder - their opponents to a degree analogous to that of any other absolutist ideology. The real challenge, currently beyond the human mindset, is to dispense with absolutism in favor of more tolerant and reason-based Enlightenment values. Proponents of this informed vision aspire to a morality well above the low bar set by Christianity. Diagram 6.1 portrays the creationist workflow for the *moral supremacy* fallacy.

Science Responds

The previous section affirms three sources of religiously-inspired violence: scripture, monotheism, and *I-do-this-in-the-name-of-god* despotism. Psychology puts forward a fourth origin, that of group dynamics, a field set on a scientific footing by the German psychologist Kurt Lewin (1890-1947).

Group dynamics studies the formation of groups and group behavior. In simple terms, groups form when individuals share common interests, experiences, or values. A psychological bond develops among members of the group, leading to behaviors that sometimes differ, both in character and intensity, from how the same individuals would act if outside the group. These in-group behaviors, which nearly universally promote the interests of the group, frequently exhibit positive characteristics. Yet the negative characteristics are more pertinent to our narrative.

Leadership and interpersonal dynamics often compel in-group members to feel superior to non-members of the group (out-group members), deprecating the status and value of the latter. In-group members might even perceive out-group members as a threat. When this occurs, it becomes simpler for in-group members to dehumanize and mistreat out-group members. This can occur in any setting - in sports, politics, professions, school cliques - and can lead to racism, sexism, and other forms of discrimination directed at out-group members.

Religion sometimes is singled out as a special case of group dynamics, owing to the intense emotional connection it inspires among its followers. This argument asserts that religion elicits a greater bond of trust and sense of community among its members, traits which shape the boundaries and interactions among the in-group. This bond, however, does not extend to the out-group, the non-believers. To promote and defend the in-group, its members can be persuaded to view non-believers as less vested in the goals of the in-group, less worthy of fair treatment, and less deserving of protection.[106] Negative labels (infidel, barbarian) and attributes (dirty, unintelligent) are ascribed to out-group members, further denigrating their status and value. In this manner, religion can inspirit its purblind followers to fanatical hatred and ruthless acts against out-group members. The absolutist nature of many religious ideologies magnifies the danger.

A related topic warrants a brief commentary. Reason and science contend that religion is not a prerequisite for positive moral behavior. Societies can, and do, develop equally effective moral philosophies without appeal to religion. Thomas Hobbes (1588-1679), John Locke (1632- 1704), Jean-Jacques Rousseau (1712-1778), and Jeremy Benthem (1748-1832) are four of the philosophers who provided a scholarly, reasoned basis for a social contract, a type of societal agreement encompassing the authority of rulers, the rights of the governed, the education of citizens, and the organizing rules of society.

Reason argues that secular education, both formal and informal, capably imparts morals. Leaders can, and do, reinforce nonreligious moral principles through word and deed. So, too, do community traditions and publicized exploits of notable, desired behaviors. The outcome is no societal anomie, no dystopia of moral chaos. Secular societies, built on a shared humanity, develop moral philosophies that are indistinguishable from religious ones: be loyal rather than disloyal, aid rather than harm, understand rather than judge, love rather than hate. Individuals in societies that subscribe to these principles live equally valid moral lives. Indeed, the secular individual is every bit as moral and principled as the pious - often more so.

MISCELLANY

Accord praise where it is due. In recent decades various popes and cardinals have issued apologies for past sins of the Catholic Church. The confessions acknowledge the violence undertaken by more than a few of its members "in service of the truth." Among the attempts at atonement, the penitents affirmed the Church's forced conversion of native peoples and its injustices against indigenous societies.[107] Its hostile acts against followers of other religions.[108] Its maltreatment and denigration of women.[109] Its condemnation of Galileo.[110] Its persecution of heretics and its violations of human rights.[111] Its inaction during the Holocaust.[112] Its separation of children from their parents.[113] Its mistreatment of gays.[114] And its cover-ups of child abuse.[115]

The creationist workflow for employing the *moral supremacy* fallacy. Diagram 6.1

The Moral Supremacy Fallacy

1. Ignore all the horrendous acts initiated by God in the Bible.
2. Ignore all the monstrous acts initiated in the name of God throughout history.
3. Ignore that atrocious acts commonly arise from absolutist ideologies.
4. Ignore that creationism is an absolutist ideology.
5. Nonetheless, utter the word Christian as a credential to declare your moral superiority.
6. Go to step 5.

[VII]

The Irrelevant Appeal Fallacy

People ache to believe that we human beings are vastly different from all other species.[116]

Daniel Dennett (1942-)

American Cognitive Scientist

Definition

The advocate of this fallacy appeals to factors not relevant to the claim. The factors typically divert attention away from unsound reasoning or want of evidence. As such, some sources refer to fallacies of *irrelevant appeal* as fallacies of distraction. Myriad creationist shibboleths expose the fallacy. One of the more common is *Jesus loves you*.

Example

Creationists routinely avow that *Christ loves you*. The Lord will bestow unconditional love, forgive every iniquity, and light the path to eternal salvation. In recompense, you need only profess unquestioned belief in His existence. Sundry scriptural attestations affirm the promise. Here is one exemplar: "We love him, because he first loved us."[117]

Recognizing The Fallacy

POINT 1. Recognize that *Jesus loves you* is but one type of irrelevant appeal. The assertion, constructed to satiate our psychological needs, adduces no evidence of a divine presence. And love, despite its critical and unarguable importance to our species, is not an indicator of external truth.

POINT 2. Recognize that the most common irrelevant appeal is *appeal to authority*, where a claim is supported not by evidence but from attribution to a figure of power or eminence. It is deference to the authority that underlies the alleged legitimacy of the claim. All such appeals, sans empirical evidence and testable predictions, are invalid. They also open the door to assorted cognitive biases, each aligned to partisan inclinations. In religion, the *appeal to authority* fallacy is closely related to the *circular reasoning* fallacy discussed in chapter four.

POINT 3. Recognize that the position, intelligence, charisma, fame, hubris, and track record of the authority are meaningless. If an assertion from authority - any authority - disagrees with experiment, it is wrong.[118] Period.

POINT 4. Recognize that a further species of irrelevant entreaty is *appeal to common sense*. This fallacy presents a claim as obvious, declaring the truth of the claim as self-evident. Readers casting a critical eye on creationist literature will be surprised at the habitual occurrence of this fallacy. Variations abound. Here are six. *It is obvious* the Noahic Flood carved the Grand Canyon. *It is against the character of God* to suggest the Bible is not His true word. *It is evident* the New Testament book of Romans adverts the Second Law of Thermodynamics.[119] *It is clear*, from God's teachings, that evolution is inconsistent with His omnipotence. *It seems reasonable* a Biblical day in Genesis bespeaks a 24-hour period. *Common sense tells us* life cannot arise on its own. All such examples, if nothing else, reinforce the adage that common sense is neither.

POINT 5. Recognize there are many types of irrelevant appeals. Most play on human emotions. Two of the more common include *appeal to the gallery* (appealing to popular sentiments) and *appeal to pity* (appealing to sympathy). The concepts of sentiment and pity are, by all means, entirely valid. Yet neither, absent of evidence, can attest to the validity of a claim.

POINT 6. Recognize that other irrelevant appeals exploit *appeal to tradition* (invoking feelings of reverence for something from the past) and *appeal to common belief* (asserting that what is believed by many must be true).

POINT 7. Recognize that still other irrelevant appeals reek of dishonesty. This is particularly true of *straw man* appeals (setting up a false argument, then attacking that argument) and *coercion* appeals (threatening an undesirable outcome).

SUMMATION. Recognize that all irrelevant appeals attempt to influence the target audience with information immaterial to the claim. That is, absent of reason or evidence. All are *non sequiturs*. All are invalid. The creationist workflow for the *irrelevant appeal* fallacy is depicted in Diagram 7.1.

Science Responds

All irrelevant appeals, when absent of evidence, are invalid. As such, science again need not respond.

MISCELLANY

There are instances, some contend, where *appeal to authority* is relevant. A commonly cited instance is when all parties to an argument agree to the expertise and credibility of the authority. This author, however, disagrees. In too many cases, acknowledgment of authority leads to erroneous results. "Authorities must prove their contentions like everybody else."[120]

The creationist workflow for employing the *irrelevant appeal* fallacy. Diagram 7.1

The Irrelevant Appeal Fallacy

1. Declare the existence, absent of empirical evidence, of the one true God.
2. To distract from absence of evidence, appeal to sentiment and sympathy.
3. To distract from absence of evidence, appeal to common sense and common beliefs.
4. To distract from absence of evidence, appeal to tradition and force.
5. Go to step 2.

[VIII]

The Post Hoc Fallacy

Science is one of the very few human activities - perhaps the only one - in which errors are systematically criticized and fairly often, in time, corrected.[121]

Karl Popper (1902-1994)
Austrian-British Philosopher

Definition

A *post hoc ergo propter hoc* fallacy (Latin for "after this, therefore because of this") is a causal fallacy. It asserts that *if B follows A, then B must have been caused by A*. The conclusion derives from the order of events, not their causality. The words *let us pray* invariably ignite the fallacy.

Example

Prayer best illustrates the *post hoc* fallacy. Petitioners who subsequently experience a favorable outcome often credit the power of prayer. Personal attestations of its efficacy abound. Consider a typical example: "A tough school exam lay ahead. I studied and studied and studied, but I feared I would fail the test. I was overwhelmed with stress. So I prayed to God, petitioning His assistance. I asked Him to walk with me, to provide me with strength and assurance. I asked Him to cast off the despair Satan had placed upon me. Almost immediately, I began to feel His love and encouragement. From His inspiration, I regained my confidence. The next day, I took the test, and I passed!"

Scriptural passages affirm the power of prayer. "And this is the confidence that we have in him, that, if we ask any thing according to his will, he heareth us."[122]

Recognizing The Fallacy

POINT 1. Recognize that inferring causality solely from chronological sequence is a logical fallacy. If it were not, anyone could claim, at whim, that a subsequent event was brought about by an ill-defined, earlier act.

POINT 2. Recognize that believers routinely cite the "evidence" of prayer predicated on chronological order alone. Yet sequence devoid of substantiating evidence cannot assert cause.

POINT 3. Recognize that no evidence exists to affirm the power of prayer.

POINT 4. Recognize that some evidence exists for the ineffectiveness of prayer.

POINT 5. Recognize that, to thwart inquiry, creationists assign theurgy to the realm of the supernatural. They then assert that science is unable to explore the supernatural; it can only study the natural world. This is a distinction without substance. It is only religion that differentiates between the natural and supernatural world. To science, it all is the natural world, and it all is discoverable.

SUMMATION. Asserting causality without evidence is meaningless. Religion is unable to substantiate that prayer is anything more than wishful thinking. Diagram 8.1 outlines the creationist workflow for the *post hoc* fallacy.

Science Responds

Scientific experiments that study the efficacy of prayer appear to be small in number. One of the more quoted studies, however, was published in the *American Heart Journal* (AHJ) in 2005.[123] The study began with 1800 patients scheduled to receive a coronary artery bypass graft. The patients were divided into three groups.

- Patients in the first group received intercessory prayer. They were informed in advance that they might or might not receive prayer.
- Patients in the second group received no intercessory prayer. They likewise were told beforehand that they might or might not receive prayer.
- Patients in the third group received intercessory prayer. They were explicitly informed in advance that they would receive prayer.

Prayer commenced the night before surgery and continued for fourteen days. The effectiveness of prayer was measured by the presence of complications within thirty days of the surgery. The result? Overall, prayer had no effect on complication-free recovery. One group, however, did experience a slightly higher degree of complications: the patients in the third group - those who received prayer and knew full well they would receive it.

The study of prayer in well-controlled trials is in its infancy. Consider some of the many variables involved. Were all patients equal in health prior to the surgery? Were some surgeries more complicated than others? Were different drugs administered before, during, or after the surgeries? Was each surgical team equally experienced? Did the patients approach the surgery with a similar mental attitude? Did all patients possess equal faith in the supernatural? Were all the prayers expressed with equal passion and conviction?

Note, too, that any benefit from prayer might derive from at least four factors. Perhaps prayer acted as a placebo. Perhaps it calmed the patient, lowering their pulse or breathing rate. Perhaps it induced a positive mental sense regarding the outcome of the surgery. Or perhaps it inspired a supernatural entity to intercede in the outcome. It is only this fourth factor, of course, that would resolve the efficacy of prayer from a religious viewpoint.

Theologians, unsurprisingly, offered their critique of the AHJ study. Three of their criticisms follow.[124] First, they noted that God responds to prayers at His will, not from the will of the petitioner. Second, God is not swayed by the quantity of prayers. And third, a simple yes-no measurement of surgical complications might disguise other meaningful responses. For example, the deity might have answered a prayer saying, in essence, "I will help you a bit" or "I will help you, but not now." Such responses would not emerge from the measurement methodology. Theologians, true to form, affirm the power of prayer from scripture: "And the prayer of faith shall save the sick, and the Lord shall raise him up."[125]

Clearly, there are difficulties. But what if the complexities could be solved? Envision a randomized, double-blind experiment where one group of people prayed for a predefined outcome, say, fewer injuries precipitated by automobile collisions in a well-defined geography. A separate contingent could serve as the control group, uttering no prayers in a distinct geography. Repeat the experiment scores of times around the world, assessing the consequences by measuring auto casualties in the target geographies. A statistically significant decline in casualties could evince the power of prayer.

If the intricacies of such an experiment could be resolved, would religion agree to conduct the trial? More important, if the results demonstrated no meaningful decline in casualties, would religion accept the meaninglessness of prayer? Science contends the answer to both questions is no. The risk to religion is too high, as a null result would verify the placebo effect of prayer, reaffirming its role as wishful thinking with no consequence on the external world.[126]

The creationist workflow for employing the *post hoc* fallacy. **Diagram 8.1**

The Post Hoc Fallacy

① Say a prayer.

② Await the desired outcome.

③ Impute a positive outcome to the power of prayer.

④ Impute a negative outcome to the wisdom of the Lord, who understands what you need better than you do.

⑤ Go to step 1.

[IX]

The Wishful Thinking Fallacy

Orgel's Second Rule

Evolution is a lot cleverer than you are.[127]

Leslie Orgel (1927-2007)

British Chemist

Definition

The protagonist of this fallacy so strongly wants something to be true, he simply declares it true. Multiple catchphrases trigger the fallacy. One of the more prevalent is *it must be true that* ...

Example

The Bible asserts all life sprang from an act of God. One such passage proclaims, "And God said, Let the earth bring forth the living creature after his kind, cattle, and creeping thing, and beast of the earth after his kind: and it was so."[128] Creationists, anxious to bear witness to scriptural truth, declare that, yes, *it must be true that* all life emanated from God. The dictum liberates uncritical followers from all inconvenient and unwanted evidence. In this instance, the religious affirmation grants a license to denounce the theory of natural selection.

Catholic creationists, keen to repudiate the theory, turn to papal authority. Pope John Paul II (1920-2005) provided such guidance in his message to the Pontifical Academy of Sciences in 1996. His speech began well. He first affirmed that no conflict exists between evolution and the doctrine of the faith.[129] He even remarked that "significant argument" supports the theory.[130] He then lost steam, concluding that, with regard to the human spirit, evolution is incompatible with the truth about man.[131] Protestant creationists articulate like-minded views.

Recognizing The Fallacy

POINT 1. Recognize there is much to be said on this topic. Bear with me.

POINT 2. Recognize that, solely from its desire to believe, creationism asserts that all life arose from God.

POINT 3. Recognize that creationists, anxious to shield their claim, once again turn to pseudoscientific alchemy, setting forth a number of assertions. Here are four. Life requires a minimum number of components to function, hence could not arise on its own. The living world portrays an astounding level of design complexity, thus requiring a designer. Many organs are too complex to arise from evolution (the human eye,[132] a dolphin's sonar melon,[133] the motorized rotating flagella of bacteria[134]). It is only from the instantaneous creation of all indispensable parts that life can emerge.[135] Recognize further that each creationist assertion, yet again, is only an assertion. Each is a fanciful canard conceived to ward off scientific heresies. Not one is supported by robust evidence.

POINT 4. Recognize that creationists, unable to substantiate their own visceral claims, choose instead to attack natural selection.

POINT 5. Recognize that, according to natural selection, species arise not from divine creation, but from random mutation and non-random adaptation. Living organisms replicate themselves but, notably, with imperfect accuracy. This leads to variation in heritable traits among offspring in a population. Some traits provide organisms with a better ability to survive in their environments; other traits, less so. The better-adapted organisms, those possessing well-suited machinery for their limited niches, survive and reproduce. Over time, their genes become more widely reflected in the population. This process leads to the rise of new species and the extinction of others. This is evolution by natural selection.

POINT 6. Recognize that the consequences of Darwin's model immediately became clear to theologians. The model requires no deity. It dispels the notion of purposeful design. It refutes the idea of goal-directed life, that is, of a theistic purpose for nature. It ousts humanity from its privileged position as an intentionally designed, essential species. It indicates we are the product of an unintelligent process - a complex and iterative response to our environment through a succession of adaptations, with no divine design or guidance. It asserts that natural selection, resulting in species that are well enough but not perfectly adapted to their habitats, produces the boundless complexity and diversity of life. It also inverts a religious archetype, where a higher intelligence creates life forms of lesser intelligence. Darwin demonstrated that natural selection is a non-intelligent process that gives rise to life forms of higher intelligence. The culmination of this process (so far) is the self-aware, contemplative, intelligent species named *H. sapiens*.

POINT 7. Recognize that creationist attacks, though many, collapse down to two fundamental arguments: that natural selection offers no testable conditions and, likewise, offers no falsifiable predictions that could disprove the theory. Both assertions could hardly be less true. The following six cases refute the attacks.

Case One Recognize that natural selection predicts a change in inherited characteristics over generations. This has been confirmed from experiment by studying 67,000 generations of the bacterium *Escherichia coli* over a 30-year period.[136] Among the findings, some bacteria were better able to exploit glucose. These organisms increased in body size, reproduced faster, and dominated the population. Some bacteria also evolved the capacity to use citrate as a new food source, a facility traced back to changes at the genetic level. Note that religion, for most of human history and still in Darwin's day, asserted that each life form was perfect in its design, unchanging from generation to generation. The *E. coli* confirmation of natural selection disproves the creationist narrative.

Case Two Recognize that natural selection predicts the emergence of new traits. This has been confirmed in the span of our lifetimes: in bacterial resistance to antibiotics; in the ability to digest milk in adulthood, a capability still spreading worldwide; in the transformation of HIV into a virus deadly to *H. sapiens*;[137] in the evolution of saltwater trout to live in fresh water;[138] in the colorization of guppies in mountain streams in Trinidad, Tobago, and Venezuela;[139] in the adaptive response of finches to shrinking food resources on the Galapagos islands;[140] in 120+ years of research on the oil content in maize at the Illinois Experimental Station in the USA;[141] and in the domestication of wild foxes in Russia[142] - to cite a handful of many, many such examples.

Case Three Recognize that natural selection predicts the stratification of fossils in accordance with the emergence and extinction of life forms. To date, paleontology confirms this prediction, concurrently disproving creationist assertions that humans and dinosaurs co-existed. Their fossils are not intermingled in the fossil record.

Case Four Recognize that natural selection predicts the existence of intermediate fossils connecting the evolution of ancient life forms to their modern descendants. Cladistics and genetics have confirmed this claim multiple times - from mosses to vascular plants; from tiny horses to present day steeds; from feathered dinosaurs to modern birds;[133] and from ungulates to whales,[144] whose evolutionary chronology from land to aquatic animal has been traced back fifty million years through at least five intermediate species. Note that creationists, despite these and analogous findings from the last few decades, perpetuate the pretense that no intermediate fossils exist. This is untrue.

Case Five Recognize that natural selection requires long periods of time which, in turn, demands an old earth. Earth is several billion years old. See the *one percent* fallacy.

Case Six Recognize that natural selection predicts a general pattern of resemblance linking together all life forms. In other words, that all living organisms arose from one or a small set of common ancestors. Genetics confirms this prediction. We all share a common DNA code. The number and configuration of the letters within the code direct (via several intermediate steps) the growth and traits of all organisms. Within each organism, it is the gene, a segment of DNA, that forms the molecular unit of heredity. The number of genes, the number of letters within each gene, and the ordering of the letters vary by life form, but the code itself is the same among all organisms.

POINT 8. Recognize that every one of the above predictions from natural selection was, and still is, falsifiable. Consider this set of potential, "what if" outcomes. The 67,000 generations of *E. coli* revealed no inherited characteristics. The studies of lizards, guppies, and foxes disclosed no emergence of new traits. Fossil layers contained an intermingling of trilobite, dinosaur, and human fossils. One mammalian fossil - yes, just one - was discovered in a geological stratum older than the first mammals. Organisms were discovered with dramatically different genetic coding processes. Earth is proved to be young, not old. Any of the preceding outcomes would strike a body blow against natural selection. But not one has occurred.

POINT 9. Recognize that in light of repeated confirmations of natural selection, creationism evolves its attacks. Unable to refute evidence of evolutionary change at molecular and cellular levels, creationism adopts a new tack. It now accepts, what it calls, micro-evolution. At the same time, it continues to deny macro-evolution (changes in larger organisms). The micro-macro demarcation allows creationists to assert the continued role of a deity in the creation of larger life forms.[145] Note, however, biology recognizes no such micro-macro boundary. The distinction is merely a creationist artifice, as untrue in Darwin's day as it is today. Darwin's own work on artificial selection demonstrated its falsity.

POINT 10. Recognize that natural selection is one of the most widely substantiated theories in all of science. It is supported from the composition and consistency of geological strata worldwide; from the ages of the strata and the fossils they contain; from the paleomagnetic alignment of the fossil record to plate tectonics and seafloor spreading; from the transitional fossils of horses, cetaceans, and other life forms; from comparative anatomy informed by cladistics; from genetic similarities and disparities within populations; from the common ancestry of hominins; from observed mutations within genes; from the structure and function of molecules in cells; and from the universal DNA code.

POINT 11. Recognize that the theory of natural selection is highly problematic for creationists. Perhaps no other scientific theory so threatens their dogmatic edifice. As such, they must continue their counterfactual onslaughts, often confusing the fact of evolution with the theory of natural selection. But remember, even if creationists succeeded in refuting every single tenet of natural selection, no such disproof would validate the creationist theory of life (see the *negative proof* fallacy in the next chapter).

SUMMATION. Recognize that the theory of natural selection is anchored by thousands of threads of evidence, not chained by one singular fact. Those awaiting - hope against hope - for a single, glorious discovery that will upend the theory should prepare themselves for disappointment. As the cognitive scientist Daniel Dennett observed, "In due course, the Darwinian Revolution will come to occupy a ... secure and untroubled place in the minds - and hearts - of every educated person on the globe."[146] Diagram 9.2 depicts the creationist workflow for the *wishful thinking* fallacy.

Science Responds

The evidence for natural selection far exceeds the cited examples. And it expands every day as new experimental results come to light. Yet creationists endure in their attacks, seeking any conceivable inroad. One approach not yet mentioned is to exploit disagreements among evolutionary biologists. Creationists leverage the discord as "proof" that scientists increasingly doubt evolution.

It is true that disagreements oftentimes arise. But the discord focuses not on the validity of the theory but its minutiae. Scientific debates, to name but a few examples, centre on evolutionary rates of change; the mechanisms behind speciation; the effects of fast breeders versus slow breeders; the sequence of ape-like and human-like features in fossil layers; the autonomous evolution of the eye across various fauna; the degree of variation in communities of reproductively isolated populations; the number of distinct hominin species; and the evolutionary significance of horizontal gene transfer. The debates frequently launch new hypotheses, each tested against physical evidence, each yielding new results, each refining Darwin's theory.

In another attack, creationists assert a form of biogenesis, the concept that life can arise only from other life. Once again, they offer no positive evidence for the claim, stating only *it must be true*. From this premise, they reject abiogenesis, the idea that life can spring from non-living matter.[147]

It is true that science cannot yet explain the first origin of life, but researchers are exploring distinct avenues: the possibility of microbial life originating below the surface of the earth; of lipids (fats, waxes) coalescing into the first self-replicating life forms; of fast tidal cycles concentrating molecules into pools rich in protonucleic acids;[148] of specific clays acting as catalysts in molecular bonding; of the self-sustaining decay of radioactive elements providing the energy to hammer out life's building blocks; and of life arising in iron- and sulphide-rich hydrothermal vents on the seabed.

Science possesses both the will and tools to explore the origin of life, as well as its continued evolution. Here again, it sports no immutable canon, no dogmatism of ignorance, no "trust me, it's true" revelations. It neither reifies ethereal objects nor fabricates vaporous spirits to explain away puzzling phenomena. Its body of knowledge is open to critique and revision, requiring no authority to authenticate its claims. Creationists can make no such statements.

It has often and confidently been asserted,
that man's origin can never be known,
but ignorance more frequently begets confidence
than does knowledge: it is those who know little,
and not those who know much, who so positively assert that
this or that problem will never be solved by science.[149]
Charles Darwin (1809-1882)
English Naturalist

MISCELLANY

Vertical gene transfer (VGT) refers to the transmission of DNA from parent to offspring. Horizontal gene transfer (HGT), in contrast, refers to transmission from non-parental sources. In recent decades, researchers have identified several types of HGT. One is the uptake of naked DNA, where a cell absorbs free-floating genetic material from the rupture of a nearby cell. Another is viral transfer, where a virus picks up DNA in one cell, then deposits it in another. HGT can occur both within and across species boundaries.

From this brief synopsis, note three points. First, recall the convenient "tree of life" metaphor, where distinct groups of flora and fauna populate each limb of the tree. Cladistics and genetics inform the ancestor and descendant relationships among the limbs. VGT and HGT, together, suggest a more tangled metaphor. Life is less a tree and more a collection of vines, each twisting upwards, downwards, and sideways. Second, the vine metaphor suggests life arose not from a single - or a small number of - ancient and distinct microorganisms, but from a messier conglomeration of primitive cells. And third, with regard to natural selection, HGT represents a source of genetic variation that is subject to the same environmental pressures as VGT. See Diagram 9.1.

The Three Domains Of Life

Diagram 9.1 Metaphors for the evolution of life.

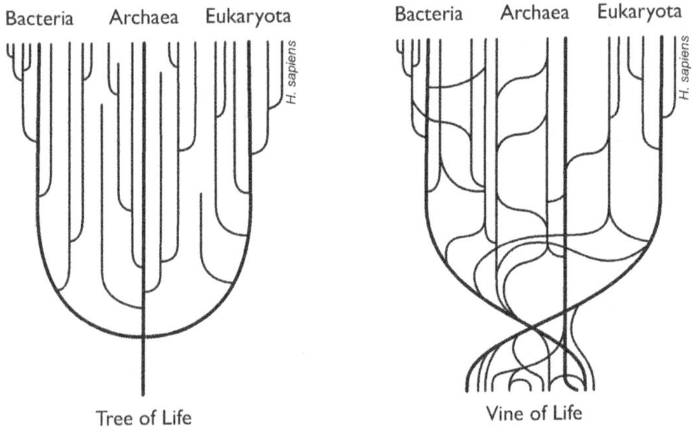

The creationist workflow for employing the *wishful thinking* fallacy. **Diagram 9.2**

The Wishful Thinking Fallacy

1. Select a claim you desperately wish to be true.
2. Assert the claim is true, devoid of substantiating evidence.
3. Vociferously attack any facts that contradict your claim.
4. Routinely reassert the truth of your claim and the falsehood of competing theories.
5. Go to step 2.

[X]

The Negative Proof Fallacy

We ought, in every instance, to submit our reasoning to the test of experiment, and never to search for truth but by the natural road of experiment and observation.[150]

Antoine Lavoisier (1743-1794)

French Chemist

Definition

The *negative proof* fallacy asserts a claim is valid because it has not been proved false. One expression that habitually evokes the fallacy is *science cannot beyond all doubt disprove that ...*

Example

Creationism asserts that because *science cannot beyond all doubt disprove* the existence of God, the deity must exist. As proof, creationists yet again cite scripture: "I am the Lord thy God."[151]

Recognizing The Fallacy

POINT 1. Recognize that absence of proof against a claim in no way constitutes proof for the claim. If such absences could be construed as positive proof, then any wild allegation could be declared true. One could claim, for example, that no evidence disproves the existence of tiny winged ponies, flying in their millions through the atmosphere of some distant star. Therefore, such creatures must exist.

POINT 2. Recognize that a valid claim must be supported by positive, confirmatory evidence. The onus rests on the claimant, not on the skeptic, to supply the evidence. The skeptic might be able to disprove a claim, but it is not her responsibility to do so.

POINT 3. Recognize that religion typically advances three arguments for the existence of God. The first is the teleological argument, which asserts that our world exhibits endless signs of complexity, elegance, and purpose. Ergo, there must be a designer, a deity as the creator of all. The teleological argument, however, follows the familiar practice of religion by offering no convincing evidence, no falsifiable predictions, and no independently verifiable conclusions.

POINT 4. The second argument is the cosmological argument. It asserts that everything has a cause. Tracing backward from cause to cause, precluding infinite regress, leads to a first cause. That first cause is God. The cosmological argument, nonetheless, is self-refuting. It argues for a deity as first cause, but does not challenge the cause of the deity, ignoring its basic tenet that everything has a cause.

POINT 5. The third argument is the ontological argument. This argument renders no comment about the universe or its contents, but affirms a connection between ideas and underlying reality. An understanding of God as the apotheosis of the most perfect conception thus bears witness to his existence in both the spiritual and temporal worlds. A common rebuttal at once presents itself, one that emanates from the core premise of the argument. If an idea implies a reality, then other "most perfect" notions must perforce exist: the myriad deities avowed by other religions, secular utopias, and similar chimeras. Further, the ontological argument is non-falsifiable, ergo nugatory.

POINT 6. Recognize that the preceding arguments, each as true to the creationist as false to the skeptic, disregard alternate explanations of observable, natural phenomena. For example, they dismiss abduction, the selection of the hypothesis that best fits empirical evidence.

POINT 7. Recognize there is a method by which religious apologists can demonstrate the existence of a supreme being. Provide evidence consistent with the scientific method that a deity exists. Provide evidence their deity, and their deity alone, created the universe. Provide evidence that explains why their deity forged the universe 13 billion years ago, not earlier or later. Provide evidence their deity is a loving, caring deity rather than a remote and indifferent deity. Provide evidence their deity is relevant, while the thousands of deities from other belief systems are irrelevant. Provide evidence their deity is indeed a deity, not merely an advanced life form that mimics the prowess of a deity. Creationists, of course, are incapable of supplying any such evidence. Equally, as noted earlier, religion willfully frames its beliefs and doctrines to circumvent empiricism, proclaiming faith as superior to observation and experiment. It is a clever, circular argument.

SUMMATION. Recognize that claims based on *negative proof* fallacies are invalid as, by definition, they provide no probative evidence. They merely attempt to shift the burden of proof. Note, too, that the more extraordinary the claim, the more extraordinary is the requisite evidence.[152] There is no more remarkable claim than the existence of a supernatural being. Hence, there exists no higher bar of compulsory evidence, yet religion offers none. Diagram 10.1 sets out the creationist workflow for the *negative proof* fallacy.

Science Responds

The onus rests on religion to provide positive evidence of its extraordinary claim. Science will respond if and when the evidence surfaces.

In the meantime, note the following. In cases where science is able to explain phenomena, such as the complexity and diversity of life on Earth, appending a deity to the equation offers no additional explanatory power. And in spheres where science cannot yet provide an explanation, such as the origin of consciousness, invoking a deity similarly yields no meaningful insights.

Worse still, attribution of phenomena to deistic authority often serves to terminate further inquiry. The practice instills an indolent mindset, one that seeks no answers beyond the realm of faith. If one believes - that is, is taught to believe - that all truth derives from a deity, that all inquiries can be answered by a banal *Goddidit*, there simply is no need to search elsewhere.

The bottom line is this: Religion claims science cannot disprove the existence of God while, in parallel, contriving every stratagem to ensure such existence is untestable.

MISCELLANY

Negative proof fallacies, sometimes referred to as *arguments from ignorance*, can be expressed in two diametric ways. They can assert a proposition is true because it has not been proved false, or they can declare an assertion is false because it has not been proved true. In both instances, the proponent interprets lack of evidence *against* his position as evidence *for* his position. This interpretation, anchored from ignorance rather than reason, often fails to distinguish *absence of evidence* from *evidence of absence*.

Why did God elect to create the universe no *earlier* than 13 billion years ago? Creationists tender two responses to this question. The first labels creation a mysterious act of the Lord, one that lies beyond human comprehension. Mere mortals, if they wish to satisfy their curiosity on this topic, must await future revelation from the deity. The second response asserts that God transcends time, as does the act of creation. Thus, there is no *earlier* moment. Note that both responses cleverly transform testable premises into non-testable splendors. Compare this behavior to that of science. As outlined in the *point in time* fallacy, science cannot yet explain the origin of the universe. Decades of research lie ahead, no doubt marked by a surfeit of false starts and frustrating cul-de-sacs. But, rest assured, when a robust explanation emerges, it will be built on testable and falsifiable predictions - never on the enigmatic act of a mythological entity.

Consider another artifice to impede testability. Conservative religions often label as blasphemous any critique of their deity's acts or existence, a stance that - both deliberately and explicitly - thwarts open inquiry and hobbles critical analysis. This conduct signals that neither the faith nor the deity can withstand impious criticism. Yet this behavior is odd, as the same religions routinely ascribe omniscience, omnipotence, and omnipresence to their deities. One wonders whether the gods, despite their talents, suffer from some form of psychological insecurity, one that forbids external examination and judgment. Perhaps this explains why so many deities demand unending praise from their subjects, a phenomenon not unlike that practiced by more than one temporal despot.

The creationist workflow for employing the *negative proof* fallacy. Diagram 10.1

The Negative Proof Fallacy

1. Assert the existence of a deity.
2. Craft the deity and its scriptures to impede testability.
3. Assert the same non-testability as a testament to the existence of the deity.
4. Fault science for its inability to disprove the existence of the deity.
5. Go to step 4.

Epilogue

What would it take to change your mind?
Response from science: *Evidence.*
Response from creationism: *Nothing.*[153]
Bill Nye (1955-) Science Educator & Ken Ham (1951-) Creationist

Formal and informal reasoning readily lend themselves to the identification and refutation of logical fallacies. Recall, though, there are instances when one form of logic - either formal or informal - declares a claim valid, while the other does not. In such cases, evidence adjudicates. Evidence stands both as judge and jurist. The British philosopher, David Hume (1711-1776), affirmed this stance in his critique of abstract logic, "If we reason *a priori*, anything may appear able to produce anything."[154]

The primacy of evidence begs a fundamental question, one worth reiterating. What is meant by evidence? Is it merely a subjective construct? Is it purely what the ardent proponent of a claim wishes it to be? Is it solely what causes one to feel good? No, far from it. It is the output of the scientific method. It is the consequence of well-formulated hypotheses empirically tested against falsifiable predictions, each conclusion dispassionately evaluated and independently verified. It is evidence forged in this manner - diligently, meticulously, and impartially - that substantiates a claim. And, to reprise an earlier point, it is the claimant who must furnish the requisite evidence.

Now, let's venture further. As noted in the *negative proof* fallacy, extraordinary claims require extraordinary evidence. What does extraordinary evidence look like? It must satisfy four tests.

PREPONDERANCE

There is no single event, no single prediction, no single moment in time that can substantiate an extraordinary claim. To satisfy the test of preponderance, a wide variety of hypotheses must examine the claim. Each hypothesis must be attended by a rigorous testing framework, established in advance of experiment. The empirical evidence must be compiled from a broad set of diverse and nonpartisan researchers. And, as always, there must be a reliable set of testable and falsifiable predictions, each to be unequivocally confirmed or refuted.

NOVELTY

To meet the test of novelty, a hypothesis must predict novel phenomena, not merely elucidate existing phenomena. It must generate predictions that few people, if any, had previously envisaged. And, ideally, it must predict outcomes that cannot be tested by available technology. This ensures that no one could furtively conduct a test, confirm an outcome, then tout an after-the-fact prediction.

METHOD

Here, science and religion diverge. Science derives its truths from the scientific method, from a way of thinking that demands critical inquiry, the formulation of testable hypotheses, the establishment of robust testing frameworks, the collection of empirical data, and the unbiased analysis of results. In competent and nonpartisan hands, all lead to the confirmation or refutation of well-constructed hypotheses. Religion, by contrast, claims the efficacy of faith-based tactics, those that accredit divine pronouncements, sacred writings, and privy attestations, all derived from theistic authority. To demonstrate the legitimacy of its philosophy, religion must address preponderance and novelty in like manner, formulating its predictions from its faith-based method of thinking. It cannot simply borrow techniques from the scientific method. The reason is plain. If religion can formulate testable predictions only from the scientific method, it is an admission its own faith-based approach is both impotent and irrelevant.

INTEGRITY

Let's presuppose religion satisfies the first three tests. It produces a voluminous catalog of novel discoveries from its faith-based thinking process, replete with detailed explanations and testable predictions. Now what? If the predictions are confirmed, the onus shifts to science to entertain the religious claims. Science should acknowledge that faith-based evidence discloses a valid theological framework for explaining the world we inhabit. But a second corollary is equally pertinent. What if the predictions are disproved? If unbiased experiments disprove faith-based claims, religion must acknowledge the findings, cease its claims, and reject its dogma. This is daunting, as religion on every occasion invents an "out" to alter, discredit, reinterpret, or otherwise dismiss unwelcome evidence. Of the four tests, the challenge of integrity is the most exacting.

Preponderance, novelty, method, and integrity. These four elements define the lofty bar of evidence demanded for extraordinary claims, a bar wholly embraced by science. No element is arbitrary. No component is subjective. No detail is dispensable. Many scientific theories have attained the bar: Newton's laws of gravity, Einstein's theories of gravitation, and Darwin's theory of natural selection. Others sought the apogee, but stumbled from disproof of their testable predictions: Ptolemy's earth-centered universe[155], Aristotle's theory of spontaneous generation[156], and Schiaparelli's discovery of Martian canals. Notice, however, these last are not failures. They merely instantiate the self-correcting nature of science - its endless inquiry, its ceaseless challenge, and its demand for testable predictions. Every claim revealed to be false narrows the venues where truth might be found. Such disproofs metamorphose into failures only when embraced as absolute, unquestionable, and divine truths. Here, queue religion.

No creationist claim has yet touched this elevated bar of evidence. Sundry reasons explain the deficiency. Here is a recap. Creation scientists (and yes, this vernacular exposes a contradiction in terms) conduct little of their own research, choosing instead to critique anomalies from mainstream science. They mandate one, and only one, dogmatic truth, then contort facts to fit their conviction. They weave yarns from small sample sizes, from cherry-picked sets of findings, and from wishful thinking. They flaunt a supreme disdain for inconvenient facts, ridiculing or rejecting large bodies of unwelcome and distasteful evidence. They exhibit a congenital inability to forsake ingrained certitudes in the face of new, ill-favored discoveries. They structure their dogmas to impede testability, labeling any disproofs as tests of faith. Above all, their unwillingness to produce rigorous, impartially assessed evidence compels creationists to rely on a distorted form of *a priori* reasoning. One founded on fabricated accounts. One erected from pseudo-Cartesian formulations: *I think, therefore God exists*.[157] And one guided, from necessity, by myriad incarnations of fallacious reasoning.

These flaws, as noted in chapter three, are endemic. Creationism contains few of the mechanisms vital to self-correction. It resists scrutiny of its core beliefs, shunning both internal dissent and external critique. It inhabits a hollow enclosure, allowing its recited doctrines to reverberate among like-minded mendicants. It endlessly parrots its precepts, each as unquestioned as it is unquestionable, to instill acceptance. Thus, "bad' information, once embraced, is difficult to evict.

Science, too, from time to time falls into the trap of logical fallacy, but with a marked distinction: Science attempts to identify and redress false reasoning. It elevates testable predictions, supported by empirical evidence, above the cut and thrust of *a priori* and deductive argument. Science recognizes that the best-reasoned arguments collapse in the face of nullifying empirical evidence. Creationism, rejecting this reality, consigns itself to a dominion of illogical, unsubstantiated, and continually disproved claims.

Religion once exerted near monopolistic control over the five W's: what, why, where, when, and how. For most of human history, it offered the only paradigm for explaining the world we inhabit. From the dawn of the Enlightenment, however, science began to encroach on religion's cherished territory. Today, it is science, not religion, that more likely can explain what will happen, why it will happen, where it will happen, when it will happen, and how it will happen. As Boyer states, "In every instance where the Church has tried to offer its own description of the world and there was some scientific alternative on the very same topic, the latter has proved better. Every battle has been lost and conclusively so."[158]

Despite this reality, Christianity claims that it, and it alone, knows the truth - yet cannot objectively demonstrate it. When pressed, the best it can do is mechanically recite clichéd *God-didits*. Creationists might assert that all life arose from the swish of a magic wand wielded by an invisible entity. They might wish it with all their hearts. But here enters the beauty of science. It has the power to disclose not what we have been indoctrinated to believe, not what we want to believe, but what actually exists. Science, as such, reflects a way of thinking, one free from dogmatic and mystical assertions. It accords an unrivaled ability to demarcate likely truths from wishful convictions. It champions a method - one absent from the lion's share of human history and possessed by no other ideology - for reaching thoughtful, credible, and defendable conclusions. In short, it teaches us how to think. And note this consequence: In science a Muslim can verify a discovery by a Sikh, an atheist can corroborate a finding by a Jew, and a Taoist can confirm a claim by a Christian. The truths of science thus impart a universal veracity wholly unadulterated by religious affiliation, political inclination, or cultural sentiment. Religious apologists can assert no similar claim.

I close this narrative with a final trigger phrase. If you engage a creationist in a friendly debate, will you know when your arguments prevail? Yes, when you hear these words: *I will pray for you*.

MISCELLANY

A more thorough examination of the four requisite elements of extraordinary evidence, complete with examples, appears in *The Dialogues*, a text by the same author of this handbook.

Schiaparelli, from his telescopic observations of Mars, reported seeing *canali* on the surface of the planet. The Italian word *canali* means both *channels* (natural structures) and *canals* (purposefully designed structures). It has been suggested that Shiaparelli meant the former, but a mistranslation of his findings into English led readers to understand the latter.[159]

Fallacies Redux

Religions are tough. Either they make no contentions which are subject to disproof or they quickly redesign doctrine after disproof. The fact that religions can be so shamelessly dishonest, so contemptuous of the intelligence of their adherents, and still flourish does not speak very well of the tough-mindedness of the believers. But it does indicate, if a demonstration were needed, that near the core of the religious experience is something remarkably resistant to rational inquiry.[160]

Carl Sagan (1934-1996)
American Physicist

Creationists employ far more fallacies than the ten cited in the main body of this handbook, each one a pretense that attempts to advocate a favored claim absent of reason and evidence. Here follows a further selection.

Ad Hominem Attacking the opponent rather than the opponent's argument. *You are not a Christian, so your understanding of Biblical truth must be limited.*

Affirming The Consequent Asserting if B follows A, then B proves A. *If God created fish, we would see fish. We see fish, so God created them.*

Appeal To Closure Demanding an answer that settles an open question, regardless of the validity of the answer (aka, not wanting to accept "I don't know" as a valid answer). *The statue wept. It must have been an act of God. There can be no other explanation.*

Apriorism Refusing to look at evidence that might counter one's claim. *I will not, and need not, read that book on evolution; it's clear God created all life.*

Argument Ad Nauseum Asserting the truth of a claim from endless repetition. *God exists.*

Argumentum Ad Populum Asserting the truth of a claim from the number of people who believe it. *More people identify themselves as Christians than any other religion. Therefore, Christianity must be true.*

Availability Bias Ascribing truth from recent, immediately available information while disregarding unwonted, harder to find facts. *Christianity decries human slavery.* [More than a few Christian denominations once considered slavery compatible with the Bible.]

Chronological Snobbery Appealing to the extensive age of something - a belief or a tradition - as proof of its truth. *The backbone of our community is, and always has been, faith; without its truths, we would be lost.*

Confirmation Bias Selecting evidence that reinforces one's ingrained beliefs. *I glanced up, just as an image of the Lord appeared in the clouds.*

Composition Asserting that the truth of a single attribute evinces the truth of the whole. *The coins mentioned in the Bible are real, historical coins; therefore, the Bible can be trusted as a true historical record.*[161]

False Analogy Asserting a claim from false and misleading comparisons. *Natural selection is equivalent to a single windstorm blowing through a junkyard and creating a 747* *ready to fly.*[162]

False Dichotomy Asserting one must choose between two, and only two, incompatible choices. *If you are not for Jesus, you must be against him*.

Insufficient Sample Drawing a conclusion from too small a sample size. *The likelihood of life arising on Earth without a creator is nil*. [Ignores that Earth is but one of likely trillions of planets, and happens to be one that hosts life.]

Non Sequitur Asserting a conclusion that does not follow from a premise. *The earthquake, clearly an overt act of God, served to punish the sinners*.

Obscurantism Opposing the discovery and spread of knowledge. *For truth, we need not look beyond the Bible*.

Reification Treating an abstract concept as a concrete object or event. *The devil made me do it*.

No True Scotsman Redirects the argument to a key term. *No "true" Christian would foment violence*.

Plurium Interrogationum Demanding a simple answer to complex questions. *God created the universe*.

Slippery Slope Asserting that a small step leads, by a chain of ambiguous events, to a dangerous culmination. *If we permit homosexuality, we place our societies at risk*.

Willful Ignorance Asserting that something cannot and should not be known. *We mere mortals cannot understand the acts of the Lord*.

CREATIONISM LAID BARE

Fallacies Critiqued

The fact that an opinion has been widely held is no evidence whatever that it is not entirely absurd.[163]
Bertrand Russell (1872-1970)
British Philosopher

Earlier chapters illustrate creationism's widespread embrace of logical fallacies. This section, from close examination of a notable creationist text, offers a more in-depth analysis. The selected text, *Scientific Creationism*, tenders a detailed comparison - one allegedly backed by factual information - of two irreconcilable models: evolution (in its broadest context) and divine creation. The volume conducts a wide-ranging assessment covering the origin of the universe, the shaping of our planet, and the rise of humanity. Its authors, many of whom possess authentic scientific credentials, proclaim it a well-researched, thoroughly-documented, authoritative tome.

Scientific Creationism was published in 1974. At first blush, it might seem odd to critique a text forty-plus years in age. Assessing a similarly-aged text from mainstream science often yields little value, as its content no longer reflects current knowledge. The same, however, is not true in creationism. Pick up a creation science text from the current decade and peruse its contents. You will discover that its arguments closely parallel those from decades earlier. This is unsurprising, as their lines of reasoning derive from the same core beliefs: from the formation of the universe and all it contains in six 24-hour days; from the generation of all life forms as functionally complete designs, evolving from no earlier organisms; from the earth's history as shaped by Noah's flood; from the creation of *H. sapiens* in the image of a deity; and from scripture as the historically accurate, inerrant word of that same deity.

Thus, the selection of a volume from 1974 is not as odd as it first might seem. Its arguments are remarkably similar to present-day tracts. A cross-check against contemporary works confirms this fact.[164] Other factors also endorse the selected text. Among them, two stand out. The subject matter is as comprehensive as any current creationist text, arguably more so. And the volume remains a widely-cited reference on creationist websites.

Now, to the fallacies. An attentive reading of *Scientific Creationism* yielded thirty-five instances of fallacious reasoning, which I cite immediately below. Their sequence matches their order of appearance in the source text. In each case, I quote the fallacy, pinpoint its type from the ten illustrated in the main body of this handbook, then provide a short commentary {in braces}. Within each quote, all *italics* are mine. Here is the list.

Fallacy #1 "It is impossible to *prove* scientifically any particular concept of origins to be true."[165] *One Percent Fallacy*. {Misunderstands the nature of science by demanding 100% proof of unwelcome claims, at the same time disregarding a preponderance of evidence against the creationist narrative.}

Fallacy #2 "This means that a philosophy of origins can only be achieved by faith, not by sight."[166] *Elevation Fallacy*. {Here, creationism confers upon itself *carte blanche* to reject any unwelcome findings from science.}

Fallacy #3 "Belief in something is necessary for true mental health."[167] *Irrelevant Appeal Fallacy*. {This statement, while arguably true, in no way attests to the specific assertions of creationism, which is but one of many possible religious and secular ideologies.}

Fallacy #4 "A life based on the whim of the moment, with no rationale, is a 'tale told by an idiot, full of sound and fury, signifying nothing'."[168] *Irrelevant Appeal Fallacy*. {Attempts, from context, to cast creationism in a more positive light by ascribing false and demeaning attributes to secularism and scientific explanations.}

Fallacy #5 "Creation is not taking place now, so far as can be determined."[169] *Negative Proof Fallacy*. {Here, the word "creation" refers to the emergence of new life forms. Even if the statement were true, it merely attacks science. It provides no positive evidence of creationism.}

Fallacy #6 "It is impossible to devise a scientific experiment to describe the creation process."[170] *Point In Time Fallacy*. {Once science is able to demonstrate abiogenesis, creationists - to rerun their fallacy - simply will seek a different unexplained phenomenon.}

Fallacy #7 "If evolution is taking place today, it operates too slowly to be measurable and, therefore, is outside the realm of empirical science."[171] *Point In Time Fallacy*. {Factually untrue. Evolutionary change has been measured, but creationism practices various dodges to dismiss the evidence.}

Fallacy #8 "Even if modern scientists should ever actually achieve the artificial creation of life from non-life, or of higher life kinds from lower kinds, in the laboratory, this would not prove in any way that such changes did, or even could, take place in the past by random processes."[172] *Negative Proof Fallacy.* {Again, even if this statement were true, it provides no positive evidence of creationism. It merely seeks to discount scientific findings.}

Fallacy #9 "Evolution cannot in any conceivable way be disproved."[173] *Wishful Thinking Fallacy*. {Factually untrue, and was so even in Darwin's era - see chapter nine.}

Fallacy #10 "Those who survive in the struggle for existence are the fittest because the fittest are the ones who survive." [174] *Wishful Thinking Fallacy*. {Attempts to transform the theory of natural selection into a tautology, ignoring the theory's testable and falsifiable conditions.}

Fallacy #11 "Evolution is an authoritarian system to be believed."[175] *Irrelevant Appeal Fallacy*. {A rich claim, emanating from arguably one of the utmost authoritarian ideologies.}

Fallacy #12 "The only reason for saying that special creation is incredible would be if one had certain knowledge that there was no God."[176] *Negative Proof Fallacy*. {Here, the words "special creation" denote the Biblical account. Note again, however, no positive evidence supports divine creation or the existence of a deity. Note, too, the tautological structure of the assertion.}

Fallacy #13 "Although many people teach evolution as though it were a proven fact of science, *it is obvious* that this is a false teaching."[177] *Circular Reasoning Fallacy*. {Declares a claim obvious, then asserts its truth from its obviousness.}

Fallacy #14 "Although many believe special creation to be an absolute fact of history, they must believe this for theological, rather than scientific reasons."[178] *Elevation Fallacy*. {Yet again, hoists faith above distasteful scientific findings.}

Fallacy #15 "Neither evolution nor creation can be either confirmed or falsified scientifically."[179] *One Percent Fallacy*. {Once again ignores the preponderance of evidence in favor of natural selection.}

Fallacy #16 "It is clear that neither evolution or creation is ... a scientific theory or a scientific hypothesis ... because neither can be tested."[180] *Wishful Thinking Fallacy*. {Factually untrue. The scientific theory offers both testable and falsifiable conditions; it only is creationism that designs its assertions to thwart testability.}

Fallacy #17 "The universe, in all its aspects evolves itself into higher levels of order."[181] *Wishful Thinking Fallacy*. {From its context in the source text, the quotation attributes a false premise to science to later launch an attack from the Second Law of Thermodynamics. Science asserts no such claim that all aspects of the universe ascend to higher levels of order.}

Fallacy #18 "Stars remain unchanged."[182] *Point In Time Fallacy*. {Factually untrue. An erroneous creationist prediction formulated decades prior to the current understanding of stellar formation, life expectancy, and collapse.}

Fallacy #19 "Creationism is consistent with the innate thoughts and daily experiences of the child."[183] *Irrelevant Appeal Fallacy*. {Religion is not innate; it is socially inculcated. As an aside, permit me to interject a dose of sarcasm. The quote from the source text might indeed be true. Creationism is consistent with a childlike mind, one drawn to imaginary creatures and engaging fables. The onus on the adult mind, however, is to practice a higher order of critical thought.}

Fallacy #20 "There is no greater stimulus to responsible behavior and earnest effort, as well as honesty and consideration for others, than the awareness that there well may be a personal Creator to whom one must give account."[184]

Moral Supremacy Fallacy. {Recites the false and arrogant claim that Christians possess higher moral values.}

Fallacy #21 "It is precisely because Biblical revelation is absolutely authoritative and perspicuous that the scientific facts, *rightly interpreted*, will give the same testimony as that of Scripture."[185] *Fallacy Of Absolutes*. {And who conducts the right interpretation?}

Fallacy #22 "If the evolutionist prefers not to believe in God, he must still believe in some kind of uncaused First Cause … But, then, who made the Matter?"[186] *Negative Proof Fallacy*. {A replay of the self-refuting cosmological argument, which argues for a deity as first cause, but does not challenge the cause of the deity.}

Fallacy #23 "The First Cause must be infinite, eternal, omnipotent, omnipresent, omniscient, moral, spiritual, volitional, truthful, loving, and living. From this, we conclude the First Cause must be a divine being."[187] *Elevation Fallacy*. {The articulated attributes just happen to correspond to Biblical assertions. A coincidence?}

Fallacy #24 "The fact that categories of natural phenomena can be arranged in orderly classification systems [such as the table of chemical elements] is a testimony to creation." [188] *Irrelevant Appeal Fallacy*. {Why is order a mark of divine creation?}

Fallacy #25 "How can one design an apparatus to observe the evolution of a star? *It is obvious* that there is no way at all by which it can be experimentally tested."[189] *Point In Time Fallacy*. {Factually untrue. Science is increasingly able to observe stellar formation in multiple galaxies.}

Fallacy #26 "The Second Law of Thermodynamics states that every system left to its own devices always tends to move from order to disorder, its energy tending to be transformed to lower levels of availability."[190] *One Percent Fallacy*. {This attack, leveled at evolution, argues that only a deity can sustain order. Yet the assault displays ignorance of the Second Law. The Law applies to isolated systems. The universe, as a whole - and as far as is known - is an isolated system. Galaxies, stars, and living organisms are not.}

Fallacy #27 "*It is obvious* by definition that neither the big-bang theory nor the steady-state theory has any observational basis."[191] *Irrelevant Appeal Fallacy*. {Once again, note the formulaic assertion: *it is obvious* that "insert any favored creationist claim here".}

Fallacy #28 "One very important unanswered question in any evolutionary model of the origin of stars and galaxies is the question of the formation of similar particles, elements, and molecules *throughout* the universe."[192] *Point In Time Fallacy*. {The Big Bang model explains the ubiquity of similar particles, elements, and molecules throughout the universe: They are carried by the expansion of the cosmos.}

Fallacy #29 "So far as we know, in fact, the solar system is quite unique in the universe. There is an almost innumerable quantity of stars, but that does not mean any of them necessarily have planets."[193] *Point In Time Fallacy*. {Factually untrue. Planets have been discovered orbiting various stars. Note: This creationist assertion is an exemplar of the *point in time* fallacy.}

Fallacy #30 "No astronomer has ever viewed any other planet outside the solar system."[194] *Point In Time Fallacy*. {Factually untrue. Many have.}

Fallacy #31 "None of the planets has any measurable amount of liquid water."[195] *Point In Time Fallacy*. {Factually untrue. At the time of this writing, water (liquid or ice) has been discovered on Mars, Earth's moon, Saturn's moon Enceladus, and Jupiter's moons Europa and Ganymede.}

Fallacy #32 "The moon and the earth have different structures and therefore different origins."[196] *Point In Time Fallacy*. {It is true the origin of the moon is not yet known. Several hypotheses suggest that a large object - a protoplanet - struck the earth in the distant past. The ejected mass formed the moon. Yet how or why potentially "different origins" favors creationist dogma is unclear.}

Fallacy #33 "The earth, with its unique hydrosphere, atmosphere, and lithosphere is, so far as all the actual evidence goes, the only body in the universe capable of sustaining higher forms of life such as man."[197] *Point In Time Fallacy*. {So far as is currently known.}

Fallacy #34 "The evolutionary model of earth history has to presume changing structures over the ages, with its physical features gradually evolving to permit the later evolution of life. There is no evidence of this, however."[198] *Point In Time Fallacy*. {Factually untrue. Enter plate tectonics.}

Fallacy #35 "In the creationist concept, man is the highest of all creatures, and thus all other created systems must in some way be oriented man-ward."[199] *Irrelevant Appeal Fallacy*. {Note the hubris. An entire universe created for a single, underserving species.}

Before I remark further on the fallacies, allow me to share several observations on methodology. My initial thought was to document every logical fallacy that appears in *Scientific Creationism*, a tome of 255 printed pages. It quickly became apparent, however, that the resulting inventory would exceed the source text in length. Such is the profusion of fallacies that crop up, on average, at a rate of two, three, or more per page. Accordingly, I was forced to truncate the exercise, curtailing the examination to the first two chapters which, together, comprise thirty-five pages. Yet, even here, nearly sixty instances of fallacious reasoning emerged. They covered the ten fallacies presented in the main body of this handbook, plus many that appear in *Fallacies Redux*. From the sixty, I selected the thirty-five that appear above, simply to match the number of pages in the first two chapters of the source text - an admittedly arbitrary device, the purpose of which was to constrain the length of this handbook.

Now, from the thirty-five fallacies and the full content of *Scientific Creationism*, I offer five summary observations.

- As noted in chapter three, *The Point In Time Fallacy*, creationist scholarship is recognized chiefly from its absence. *Scientific Creationism* confirms this phenomenon. The text purports to lay out a somewhat rational analysis of evidence but, in fact, dedicates considerable real estate to attacking findings from mainstream science, while placing its own creationist evidence - to the extent any is presented - under no similar microscope. Instead, verbiage similar to *it is obvious that* and *evidence consistent with* (consistent, not causal) repeatedly is put forward to defend the creationist stance.

- Continuing the previous point, the creationist assertions that appear in the source text are just that - assertions. Each one evidence-free. Each one partisan. Each one engineered to tout one, and only one, specific truth. There is no open inquiry. There is no formulation of hypotheses that permit dissimilar outcomes. There is no dispassionate evaluation of evidence. There is no admission of any testable condition that, if falsified, would collapse the dogmatic house of cards. In short, there is no "science" in scientific creationism. It is this reality - one that affirms a central theme of this handbook - that forces creationism to exploit every use of logical fallacy.

It has no other choice.

- The thirty-five instances cited above emphasize *point in time* fallacies. This is no accident. Science, as previously noted, is an eternally dynamic enterprise. As it advances, it becomes increasingly capable not only of disproving mystical assertions, but also of offering a more robust understanding of every aspect of the cosmos. Creationism never seems to understand this. Its proponents continue to assert - regularly and confidently - various *point in time* fallacies. They seem to remain oblivious to the fact that, with every likelihood, their testable assertions soon will be disproved. *Scientific Creationism* evinces the point, with many of its assertions from 1974 now factually refuted.

- The source text reveals further insights, the most prominent of which is this. The authors, to buttress creationist dogma, cite scripture no fewer than 270 times - a fantastic example of the extensive use of *circular reasoning* and *irrelevant appeal* (notably, *appeal to authority* and *appeal to emotion*) to assert dubious claims. Each citation serves to attack science, instill a favored interpretation, or proclaim nebulous *Goddidits*.

- In the final chapter, what advice is tendered to the overly-inquisitive mortal who foolishly might question Holy Writ? He is admonished to "bow before its Author in believing obedience."[200] Such is the nature of "science" according to creationism.

I close this section by quoting, one last time, from *Scientific Creationism*. To the proponents of creationist ideologies, "saying it does not make it so."[201]

MISCELLANY

A refutation from mainstream science of the assertions put forward in *Scientific Creationism* appears in Hughes.[202] Among the criticisms and disproofs lifted directly from the latter text, and targeted at the former, are the following: a misunderstanding of science, misrepresentation of scientific knowledge, use of out-of-date material, omission of opposing lines of evidence, absurd conclusions, inappropriate and misidentified sources, attacks on the honesty of scientists, use of discredited arguments, and silly calculations.

A Letter To Creationists

Reason conquers all.[203]
Marcus Manilius (circa 100 CE)
Roman Poet

Dear Creationists,

I am a secular humanist, one who bases his worldview on Enlightenment values. One such value is skepticism. Please note that skepticism in no way denotes robotic disbelief. The skeptic merely seeks a reasonable body of evidence as the basis for reaching reliable and valid conclusions. We skeptics, if I may speak on behalf of my like-minded brethren, will believe anything if substantiated by sound and sufficient evidence.

As a skeptic, my location on an ideological spectrum differs considerably from yours. But, on occasion, we do speak to one another. In those moments, you might wish to persuade me of your beliefs. That is fine, as long as the conversation is calm and rational. I suggest we start by acknowledging that each of us is, at the very least, somewhat intelligent and somewhat reasonable. This awareness, I'm afraid, sometimes evaporates in moments of sharp disagreement. So let us try to retain this thought at the forefront of our minds.

It also might be helpful if you learn how to talk to a skeptic. Here I imply no pejorative. I simply suggest that some techniques will stand you in better stead as you express your views. In this spirit, allow me to offer four nuggets of advice.

First, do not quote scripture. You might believe reciting scripture buttresses a claim. In fact, the opposite is true. Skeptics quickly recognize *appeals to authority* and *circular reasoning* as intellectual fragility. The invocation of such fallacies not only signals an ideology unable to withstand objective scrutiny, but semaphores a compulsion to sustain alleged truths from mandate of temporal authority. So, among all the fallacies employed by creationism, do not cite scripture.

Second, do not accuse science of indoctrination. More than one of your colleagues levels this accusation, asserting that mainstream education indoctrinates hapless souls into the scientific method. The assertion not only is untrue, it also emits the unpleasant odor of hypocrisy. Yes, the scientific method is a way of thinking, one designed to improve our comprehension of the cosmos and all it contains. But it posits no specific truth or destination. It molds no minds in alignment to a single, unquestionable belief. It avers no intransigent conviction to which all evidence must conform. On the contrary, science teaches *how to think*, not *what to think*. It is open to any hypothesis substantiated from evidence. This stands in stark contrast to religion - any religion - that extols one, and only one, dogmatic truth.

Consider this distinction. If you have attended a scientific conference (or watched one on YouTube), you will have noticed that the speaker, upon completing her dissertation, is grilled by the audience. Questioners subject the hypotheses, methodologies, predictions, and evidence to intense scrutiny - sometimes acridly so. There is no simple "well done, we believe you, let's not ask any questions." Now, compare this to a church service, where the congregation imbibes one specific truth absent of inquiry, devoid of challenge, and bereft of alternate viewpoints. Which forum reeks of greater indoctrination?

Third, accord evidence the highest respect. Evidence is the *sine qua non* to skeptics. If you wish to persuade us a claim is valid, impart evidence. Explain your hypothesis, disclose your testable predictions, and outline the conditions that, if they arise, would disprove your claim. Just as critically, do not debase, dismiss, or disparage the role of evidence. And do not proffer endless excuses for the absence of evidence. Your deity, so you allege, is omnipotent, omniscient, and omnipresent. It should be the simplest of tasks to obtain truly exemplary evidence of your claims.

And last, ready yourself to answer a single question. What would change your mind? Skeptics answer this question with a single word. Evidence. From evidence, science stands ready to revise its theories - its provisional truths - to reflect a better understanding of the world. So do skeptics. This is a frequent occurrence as fresh minds articulate new paradigms, as investigative technologies improve, and as new evidence comes to light. So be prepared with your response. And here is a hint. When the question is posed, do not say "nothing." This simply will communicate an inability to critically examine ingrained beliefs, an unwillingness to consider unwelcome information, and a deep-seated craving to remain within the inner recesses of one's echo chamber. In short, it exposes a mindset unable to disavow.

Thank you, creationists, for considering my comments. I look forward to our next conversation.

With the greatest sincerity,
Serge

Advice to creationists on how to talk to a skeptic. **Diagram 11.1**

Advice To Creationists

1. Do not cite scripture. Reliance on Holy Writ and its inherent logical fallacies signal weakness in your ideology.
2. Do not accuse science of indoctrination. Science never blindly asserts one, and only one, specific truth.
3. Substantiate extraordinary claims with extraordinary evidence.
4. Be prepared to answer this question: What would it take to alter your views?

A Letter To Charles

Every why hath a wherefore.[204]
William Shakespeare (c.1564-1616)
English Playwright

Dear Charles,

Yes, a letter has arrived. Perhaps the first in many a year. Few of your former correspondents, I suspect, elect to write. I cannot censure them, however, for I, too, have been remiss. But now I sit, pen in hand, primed to rectify my negligence.

Let me begin by assuring you that, while gone, you are not forgotten. Many honor your legacy. Your theory of natural selection, confirmed as the primary mechanism that drives evolution, stands at the apex of scientific thought. Your name - frequently uttered in the same breath as Galileo, Newton, and Einstein - connotes both genius and integrity. Your work launched revolutions in multiple disciplines, among them biology and geology, paleontology and philosophy, and psychology and religion. And for untold millions, you altered how we comprehend the world and our place in it. No, you need have no misgivings over your intellectual estate. It resides in sound hands.

A profusion of tributes further attests both to your scholarly and pragmatic heritage. Some honors, you will recall, were bestowed during your lifetime. More than a century has passed since those prolific days, so allow me to refresh your memory by citing four of the more prominent. Each recognized your outstanding contributions to science.

1839 The Royal Society, the oldest surviving scientific academy in the world, awarded you a Fellowship.[205] You had just attained the tender age of thirty.

1859 The Geological Society of London, in the same year that you published *On the Origin of Species*, conferred upon you the Wollaston Medal.[206]

1864 Five years later, the Royal Society bestowed upon you the prestigious Copley Medal.[207] You will recall that the medal, first granted in 1731, is the oldest extant scientific award in the world. You were its 134th recipient.

1877 Cambridge University awarded you an honorary Doctor of Laws degree for eminent achievement,[208] six years after the publication of your second great masterwork, *The Descent of Man*.

Now, perhaps I might surprise you. The honors, upon your demise, arrived not less frequently but more. Today, posthumous laurels appear in the eponymous appellations of countless places, organizations, organisms, and objects. Here is a sampling: the city of Darwin in Australia; Darwin Glacier in California; Darwin Island in Antartica; Darwin College in England; Darwin Sound in Canada; the Darwin crater on the moon; the Darwin asteroid between Mars and Jupiter; the frog *Ingerana charlesdarwini*; the dinosaur *Darwinsaurus*; and the Darwin-

Wallace medal, awarded by the Linnean Society to individuals who make notable contributions to evolutionary biology.[209] Apart from namesakes, a life-sized statue celebrates your genius at the Natural History Museum in London, as do sculptures elsewhere in England and beyond. Still further recognition derives from your image, which adorns the postage stamps of myriad nations, as well as the ten-pound banknote in the United Kingdom.

I know your modesty disdains fawning praise. Yet I hope this compendium affords you some small comfort. The same compilation dislodges from me an admission, one that reveals the true motive of this missive. I write you to remedy an oversight. Your register of awards, though both lengthy and well-earned, is, I believe, incomplete. I suggest it omits a citation of singular import, one I have not encountered elsewhere. To rectify the omission, I here elect to accord to you the delinquent accolade. It is this: *On the Origin of Species* is, quite possibly, the most honest book ever written. Allow me to reiterate this bold assertion. Your opus is likely the most principled text ever penned.

Such an assertion, at first sight, might sound implausible. Perhaps even biased. Let me assure you, however, the declaration arises from no partisan allegiance to your codification of natural selection. It likewise emanates from no unquestioned deference to evolutionary theory. And it emerges from no fallacy of *wishful thinking* - I do not declare the claim true simply because I wish it so. The assertion arises from none of these sources. It stems, instead, from a factual, empirical wellspring: from the criticisms that appear in your text, criticisms that you yourself cite. Permit me to explain.

You dedicate approximately sixty percent of your book to articulate your theory. You begin by describing artificial selection, proceed to natural selection, then cite example after example from assorted ecological niches to substantiate your theory. Yet it is the other forty percent that impresses me most. Here, you openly acknowledge problems with your theory. You cite numerous gaps, some raised by your opponents, others advanced by yourself. In selected cases, you cite additional evidence to address the gaps. In others, however, you admit the evidence essential to your theory does not exist. More than once, you pointedly write "fatal to the theory"[210] or words to similar effect. You accept that if supporting evidence is not found, or if contradictory evidence arises, your theory will be disproved.

Permit me again to refresh your memory. Here follow twenty-some gaps lifted directly from your text.

- Egyptian plants and animals, which have been studied for several thousand years, show no sign of evolutionary change.[211]
- Species differ from each other not in single characteristics but in a multitude of ways. Natural selection cannot explain the simultaneous evolution of so many characteristics in tandem.[212]
- Many species possess traits of little or no value, such as a giraffe's tail. Such superfluous traits cannot have been shaped by survival pressures.[213]
- Larger bodies require greater sustenance. Food is scarce in nature, so natural selection cannot account for organisms with larger body sizes.[214]
- If natural selection bestowed a beneficial trait on one species, such as the giraffe's ability to reach high food sources, why has it not endowed the same advantage on other even-toed ungulates?[215]
- Similarly, if man is related to ape, why has ape not acquired the mental powers of our species?[216]

- It is difficult to envision how an endless series of minute modifications can produce a distinct object - a maple leaf - that then reproduces itself, from tree to tree, in unvarying form.[217]
- Flounders, when young, possess two eyes, one on each side of their heads. As flounders mature, it is unclear how natural selection urges one eye to slowly and methodically migrate to the other side of their heads.[218]
- Natural selection cannot explain how some species appear "with suddenness and by modifications appearing all at once."[219]
- Species evolve not from natural selection but from an internal force.[220]
- If natural selection is true, we should see a multitude of transitional forms alive today.[221]
- If natural selection is true, we should see a clear and complete record of transitional forms in the fossil record, but we do not.[222]
- For natural selection to produce its results, the earth must be considerably older than currently known.[223]
- It is unclear how natural selection shapes behaviors, such as a bee's tendency to construct six-sided cells.[224]
- If offspring inherit a blend of features from their parents, why are we not all blended into a drab, unremarkable uniformity?[225]
- If natural selection can convert a land animal into an aquatic creature, how could transitional forms (part terrestrial, part marine) subsist?[226]
- How can natural selection create organs of extreme perfection - the eye, the electric organs of fishes - which achieve their purpose only when all of their parts exist and operate in unison?[227]
- How can natural selection explain the sheer beauty of so

many plants and animals, a beauty whose purpose is to delight *H. sapiens*?[228]
- How can natural selection account for insects who differ greatly from their parents, but who themselves are sterile and cannot bear offspring?[229]
- Natural selection cannot account for the simultaneous emergence of entire groups of species, such as those arising from the Cambrian Explosion.[230]
- Natural selection likewise cannot account for the sudden extinction of whole groups of species.[231]
- Why is there a paucity of fossils below the Cambrian strata?[232]

Here you stand, Charles, articulating what arguably is the most profound idea of your era. Yet you conjure no *negative proof* fallacy, asserting your theory is true because it has not been proved false. You evoke no *appeal to authority*, labeling your idea as the revealed wisdom of an inerrant, unseen entity. You offer up no hermeneutic parsing of a primitive, enigmatic scrawl to proclaim the verity of your thesis. You tout no *wishful thinking*, proclaiming the truth of your theory to satiate a psychological yearning. You advance no sophistry of *moral supremacy* to instill its acceptance. And, notably, you craft no excuses for essential but absent evidence.

Instead, in meticulous detail, you transparently formulate your theory. You openly acknowledge its gaps and possible refutations. You cite explicit conditions that could falsify the theory. You encourage robust inquiry and challenge. You invite independent and dispassionate verification. And you recognize that confirmation or disproof must await both scientific and technological advances. These traits, to my mind, exemplify your adherence to the highest principles of intellectual honesty, scientific rectitude, and moral integrity. It is from this reputable bedrock, this substratum of enduring probity, that I accord to you the accolade *Most Honest Book Ever Written*. There is, frankly, no higher bar.

I close this letter with three rhetorical queries. Each exemplifies your strength of character by vocalizing its antithesis. Have you encountered creationists who, if their predictions were unambiguously refuted by empirical evidence, would openly acknowledge the failure of their ideology? Can you cite a religious catechism that not only articulates its doctrines, but also openly enumerates potentially fatal gaps in its teachings? Can you name a religious tome that lays out a precise set of reasonable, testable conditions that, if they arise, might unequivocally disprove its assertions? Neither can I.

With utmost admiration,
Serge

About The Author

Critique the idea, not its proponent.
Serge Bourgaine

Extracts from two prior books, *Never By Authority* and *The Dialogues*, form the content of this handbook. The earlier works explore in considerably greater detail the intersection between science and religion. The current amalgam brings explicit focus to a single topic - the cache of fallacies that underlie creationist claims.

Among the ten fallacies presented in this handbook, which do creationists most commonly exploit? A tally places *circular reasoning* and *irrelevant appeal* at the top of the list. The *point in time* and *negative proof* fallacies emerge next most frequently. Note, however, that creationist fallacies, like quarks, rarely appear alone. Pick up any piece of creationist literature, turn to a random page, read a paragraph and, chances are, you will encounter fallacy upon fallacy. This, I suspect, is no accident. The layering of fallacies puts a superficial sheen on creationist assertions, forcing one to peel, layer by layer, through a series of specious and wishful claims. The layers, once shed, invariably expose an intellectually vacuous, inner core - one populated solely from the myths and superstitions embedded in ancient tales.

Creationists know this. They are clever polemists. The layering of fallacies, each constructed to cloak an absence of evidence, is an effective and all-too-often successful stratagem. Consider, too, the workflows that close each chapter of this handbook. Each provides tongue-in-cheek instructions for creating the respective fallacy. At times, one is tempted to think that creationists bind the various workflows into a guide for distribution among their like-minded colleagues. You be the judge.

For those who wish to condense this handbook down to a single message, it is this: When a claim - any claim - is asserted, ask the claimant to articulate the reasonable and testable conditions that, if disproved, would cause the proponent to disavow their claim. If the advocate is unable or unwilling to present such conditions, the claim has no merit.

I close with an acknowledgment lifted from earlier texts. Serge Bourgaine is a *nom de plume*. The rationale behind its use, once again, bears repeating. The author contends that no apologia should derive validity from the standing, acumen, or renown (or anonymity) of its proponent. To do so simply would invoke the *appeal to authority* fallacy, one of the many types of irrelevant appeal described in chapter seven. The facts, claims, and conclusions presented in this book thus stand on their own to be reflected upon by the reader, and to be judged solely from empirical evidence backed by testable predictions.

Serge

Endnotes

Entries that omit page numbers refer, in nearly all instances, to ebooks or similar digital content.

[1] Jefferson 1820, p. 1. Source text: *The most restive is that of the priests of the different religious sects who dread the advance of science as witches do the approach of daylight.*

[2] Aristotle 2003a, Bekker 25a1+, pp. 39+. Also Aristotle 2003c, Bekker 21a1, p. 32.

[3] Hume 1960, p. 124.

[4] Wilson 2014, not paginated. Source text: *If one does not accept the recent six-day creation understanding then that person is actually not a 'Seventh-day' Adventist since the seventh-day Sabbath would become absolutely meaningless historically and theologically and most of our Biblically-based doctrines centered in Christ and His authoritative voice would become meaningless as well.*

[5] Voltaire 1900, p. 170. Source text: *Traviallons sans raisonner, dit Martin; c'est le seul moyen de rendre la vie supportable.*

[6] King James Bible Online, Genesis 1:1.

[7] Harris 2011, timestamp 01:06:57.

[8] Dalrymple 1991, pp. 125+.

[9] King James Bible Online, Genesis 15:30; Exodus 12:40.

[10] King James Bible Online, 1 Kings 6:1-38.

[11] King James Bible Online, Genesis 5:5.

[12] Ashton 2009, p. 107.

[13] Dawkins 2009, not paginated.

[14] Dawkins 2009, not paginated.

[15] Mendez 2013, p. 454.

[16] Shoemaker 1983, p. 489.

[17] Williams 2000, p. 56.

[18] Suyu 2010, p. 219.

[19] Lunardini 1995, p. 20.

[20] Liu 2000, p. 186.

[21] Ford 1999, p. 88.

[22] Ashton 2009, p. 152.

[23] Ashton 2009, p. 323.
[24] Ashton 2009, p. 53.
[25] Tobin 2012, not paginated.
[26] Henke 2005, not paginated.
[27] Ashton 2009, p. 12.
[28] Morris 1974, p. 204.
[29] Tyson 2019, p. 25.
[30] Halley 1714, pp. 296-300.
[31] See Works Cited.
[32] Voltaire 1900, pp. 2-5. Source text: *Remarquez bien que les nez ont été faits pour porter des lunettes, aussi avons-nous des lunettes. Les jambes sont visiblement instituées pour êtres chassées.*
[33] Tegmark 2018, paraphrased from timestamp 00:42:05.
[34] King James Bible Online, John 17:17.
[35] Dawkins 2018, paraphrased from lecture.
[36] King James Bible Online, Joshua 10:12-13.
[37] Aquinas 2003, p. 177.
[38] King James Bible Online, Exodus 14:21.
[39] King James Bible Online, 2 Kings 2:8.
[40] King James Bible Online, Mark 5:11-13.
[41] King James Bible Online, Jonah 2:1-10.
[42] King James Bible Online, 2 Kings 6:5-6.
[43] Pope Gregory I 1911, pp. 232-233.
[44] Reiss 2015, pp. 1-10, 17-18, 146.
[45] Aquinas 2003, p. 350.
[46] Aquinas 2003, p. 358.
[47] Aquinas 2003, p. 296.
[48] Aquinas 2003, p. 356.
[49] Aquinas 2003, p. 359.
[50] Aquinas 2003, pp. 361-2.
[51] Aquinas 2003, p. 355.
[52] Aquinas 2003, pp. 354-5.
[53] Aquinas 2003, p. 489.
[54] Aquinas 2003, pp. 564-5.
[55] King James Bible Online, Genesis 1:3 versus Genesis 1:16.
[56] King James Bible Online, Genesis 1:11 versus Genesis 1:16.
[57] King James Bible Online, Genesis 1:1 versus Genesis 1:16.
[58] King James Bible Online, Genesis 1:11 versus Genesis 1:16.
[59] King James Bible Online, Genesis 2:5.
[60] King James Bible Online, Genesis 1:1 versus Genesis 1:16.
[61] King James Bible Online, Genesis 1:9.
[62] King James Bible Online, Genesis 1:20 versus Genesis 1:24.
[63] King James Bible Online, Genesis 1:21 versus Genesis 1:24.
[64] King James Bible Online, 1 Chronicles 16:30, Psalms 93:1, and Psalms 96:10.
[65] Walmesley 1825, pp. 139, 350. Asserted by the English Catholic Bishop

Charles Walmesley (1722-1797).
[66] King James Bible Online, Matthew 16:28; Matthew 24:34; Luke 21:32. Asserted by Jesus.
[67] Cran 1998, not paginated. Asserted by Jewish Essenes sect.
[68] Lazarus 1998, p. 237. Asserted by Martin of Tours (c.320-397).
[69] Randi 1990, p. 236. Asserted by English Cardinal John of Toledo (?-1275).
[70] Lazarus 1998, p. 237. Asserted by Italian Pope Innocent III (?-1275).
[71] Festinger 2014, not paginated. Asserted by German Anabaptist prophet Melchior Hoffmann (c.1495-c.1543).
[72] Nichol 1845, p. 12. Asserted by American Baptist lay preacher William Miller (1782-1849).
[73] Epp 2018, not paginated. Asserted by German minister Claas Epp (1838-1913).
[74] Blackstone 1916, pp. 7-8. Asserted by American Methodist minister William Blackstone (1841-1935).
[75] Armstrong 1934, p. 5. Asserted by American evangelist Herbert Armstrong (1892-1986).
[76] Smith 1978, pp. 35-36. Asserted by American pastor Charles Smith (1927-2013).
[77] Whisenant 1988, p. 1. Asserted by American Christian prognosticator Edgar Whisenant (1932-2001).
[78] Robertson 1990, pp. 312-313. Asserted by American Southern Baptist minister Pat Robertson (1930-).
[79] Barisic 1999, not paginated. Asserted by American Southern Baptist pastor Jerry Falwell (1933-2007).
[80] Horkheimer 1947, p. 24.
[81] King James Bible Online, Proverbs 30:5.
[82] King James Bible Online, 2 Samuel 7:28.
[83] King James Bible Online, 2 Timothy 3:16.
[84] Calvin 1845, p. 261.
[85] Augustine 2003, p. 46. Source text: *We are too weak to discover the truth by reason alone and for this reason need the authority of sacred books*.
[86] Aquinas 1924, p. 11. Source text: *That it is not a mark of levity to assent to the things that are of faith, although they are above reason*.
[87] Pope Pius IX 1864, error 8.
[88] Pope Pius IX 1870, session 4.
[89] Pope Benedict XVI 2009, article 56.
[90] Luther 1914, p. 126. Source text: *Ratio … ist die hochste hur, die der Teuffel hat*.
[91] King James Bible Online, Romans 14:23.
[92] King James Bible Online, Proverbs 3:5.
[93] King James Bible Online, Romans 3:28.
[94] King James Bible Online, James 2:24.
[95] King James Bible Online, Isaiah 42:5.

[96] King James Bible Online, Joshua 10:12-13.
[97] Russell 1997, paraphrased from pp. 547-8.
[98] Collins Dictionary. [collinsdictionary.com]
[99] Sagan 1996, not paginated.
[100] Russell 1950, p. 104.
[101] King James Bible Online, Matthew 22:39.
[102] King James Bible Online, Numbers 31:17-18.
[103] King James Bible Online, Luke 19:27.
[104] King James Bible Online, Matthew 28:19.
[105] Veenhoven 2007, not paginated.
[106] Teehan 2010, not paginated.
[107] Pope Apologises for Church Sins 2000, not paginated.
[108] Pope John Paul II 2000, not paginated.
[109] Pope John Paul II 1995, not paginated.
[110] Pope John Paul II 1992, not paginated.
[111] Carroll 2000, not paginated.
[112] Cassidy 1998, not paginated.
[113] Pope Sends First Email Apology 2001, not paginated.
[114] Gallagher 2016, not paginated.
[115] Pope Francis Condemns Child Sex Abuse and Church Cover-Ups 2018, not paginated.
[116] Dennett 1995, p. 338.
[117] King James Bible, 1 John 4:19.
[118] Feynman 1994, p. 150.
[119] Morris 1974, p. 212. Here, Morris interprets the words *bondage of corruption* from Romans 8:21 to mean *bondage of decay*. He then equates *decay* to *entropy*.
[120] Sagan 1996, not paginated.
[121] Popper 1965, p. 216.
[122] King James Bible Online, 1 John 5:14.
[123] Benson 2005, pp. 934-942.
[124] Jantos 2007, pp. S51-S53.
[125] King James Bible Online, James 5:15.
[126] Bragg, Science and Religion, 25 January 2001.
[127] Gregory 2007, not paginated.
[128] King James Bible Online, Genesis 1:24.
[129] Pope John Paul II 1996, article 3. Source text: *No había oposición entre la evolución y la doctrina de la fe*.
[130] Pope John Paul II 1996, article 4. Source text: *La teoría de la evolución es más que una hipótesis … La convergencia, de ningún modo buscada o provocada, de los resultados de trabajos realizados independientemente unos de otros, constituye de suyo un argumento significativo en favor de esta teoría*.
[131] Pope John Paul II 1996, article 5. Source text: *Las teorías de la evolución … son incapaces de fundar la dignidad de la persona*.
[132] Ashton 2009, p. 141.

[133] Ashton 2009, p. 79.
[134] Ashton 2009, p. 140.
[135] Ashton 2009, p. 27.
[136] Lenski 2018, not paginated.
[137] Bragg, The Origin of Infectious Diseases, 1 June 2011.
[138] Willoughby 2008, not paginated.
[139] Dawkins 2009, not paginated.
[140] Lents 2019, not paginated.
[141] Dawkins 2009, not paginated.
[142] Dawkins 2009, not paginated.
[143] Bragg, Feathered Dinosaurs, 26 October 2017.
[144] Bragg, The Whale - A History, 21 May 2009.
[145] Ashton 2009, p. 47.
[146] Dennett 1995, p. 19.
[147] Ashton 2009, pp. 40-42.
[148] Dorminey 2019, not paginated.
[149] Darwin 2003a, p. 253.
[150] Lavoisier 1793, p. x of Preface. Source text: ... *à le mettre continuellement à l'épreuve de l'expérience; à ne conserver que les faits qui ne font que des données de la nature, & qui ne peuvent nous tromper; à ne chercher la vérité que dans l'enchaînement naturel des expériences & des observations*.
[151] King James Bible Online, Exodus 20:2.
[152] Sagan 2000, timestamp 00:01:25.
[153] Nye 2014, paraphrased from timestamp 02:03:54 through 02:07:18.
[154] Hume 2003, p. 509. Note: Hume also challenged the absolute truth of a posteriori reasoning, but not its affirmation of provisional truth.
[155] Ptolemy 2003, pp. 9-12.
[156] Aristotle 2003b, Bekker 569a25, p. 95.
[157] Descartes 2003, pp. 275-276.
[158] Boyer 2001, p. 320.
[159] Washam 2010, not paginated.
[160] Sagan 1979, pp. 332-333.
[161] Ashton 2009, p. 283.
[162] Krauss 2015, paraphrased from online article.
[163] Russell 1930, p. 50. Source text: *The fact that an opinion has been widely held is no evidence whatever that it is not entirely absurd; indeed, in view of the silliness of the majority of mankind, a widespread belief is more likely to be foolish than sensible*.
[164] One such contemporary text is *In Six Days: Why Fifty Scientists Choose to Believe in Creation*. See *Works Cited*.
[165] Morris 1974, p. 4.
[166] Morris 1974, p. 4.
[167] Morris 1974, p. 5.
[168] Morris 1974, p. 5. Here, Morris quotes from Shakespeare's Macbeth.
[169] Morris 1974, p. 5.

[170] Morris 1974, p. 5.
[171] Morris 1974, p. 5.
[172] Morris 1974, p. 6.
[173] Morris 1974, p. 6.
[174] Morris 1974, p. 7.
[175] Morris 1974, p. 7.
[176] Morris 1974, p. 8.
[177] Morris 1974, p. 8.
[178] Morris 1974, p. 9.
[179] Morris 1974, p. 9.
[180] Morris 1974, p. 9.
[181] Morris 1974, p. 10.
[182] Morris 1974, pp. 13, 25.
[183] Morris 1974, p. 14.
[184] Morris 1974, p. 14.
[185] Morris 1974, p. 15.
[186] Morris 1974, p. 19.
[187] Morris 1974, paraphrased from p. 20.
[188] Morris 1974, paraphrased from pp. 21-22.
[189] Morris 1974, p. 24.
[190] Morris 1974, p. 25.
[191] Morris 1974, p. 26.
[192] Morris 1974, p. 27.
[193] Morris 1974, p. 28.
[194] Morris 1974, p. 28.
[195] Morris 1974, p. 29.
[196] Morris 1974, p. 30.
[197] Morris 1974, p. 32.
[198] Morris 1974, p. 32.
[199] Morris 1974, p. 35.
[200] Morris 1974, p. 206.
[201] Morris 1974, p. 200.
[202] Hughes 1992, pp. 97-102.
[203] Manilius 1920, p. 125. Source text: *Ratio omnia vincit*.
[204] Shakespeare, p. 153, 2.2.44.
[205] The Royal Society.
[206] The Geological Society.
[207] The Royal Society.
[208] Cambridge University 1922, p. 58.
[209] The Linnean Society.
[210] Darwin 2003b, p. 162.
[211] Darwin 2003b, p. 99.
[212] Darwin 2003b, pp. 99-100.
[213] Darwin 2003b, p. 100.
[214] Darwin 2003b, p. 104.
[215] Darwin 2003b, pp. 104-105.

[216] Darwin 2003b, p. 106.
[217] Darwin 2003b, p. 106.
[218] Darwin 2003b, pp. 108-109.
[219] Darwin 2003b, p. 117.
[220] Darwin 2003b, p. 116.
[221] Darwin 2003b, p. 80.
[222] Darwin 2003b, p. 152.
[223] Darwin 2003b, p. 153.
[224] Darwin 2003b, p. 80.
[225] Darwin 2003b, p. 152.
[226] Darwin 2003b, pp. 82-83.
[227] Darwin 2003b, p. 85, 89.
[228] Darwin 2003b, p. 94.
[229] Darwin 2003b, p. 132.
[230] Darwin 2003b, pp. 162-163.
[231] Darwin 2003b, p. 170.
[232] Darwin 2003b, p. 165.

Works Cited

The bibliography is organized by author surname and publication date.

Aquinas, Thomas. 1924. *The Summa Contra Gentiles of Saint Thomas Aquinas, First Book*. First published circa 1265. Translated by the English Dominican Fathers. London: Burns Oates & Washbourne.

Aquinas, Thomas. 2003. *The Summa Theologica, Volume 1*. First published circa 1485. Great Books of the Western World, Volume 17. Edited by Mortimer J. Adler, translated by Laurence Shapcote, seventh printing. Chicago: Encyclopædia Britannica.

Aristotle. 2003a. *Prior Analytics*. First published circa 350 BCE. Great Books of the Western World, Volume 7. Edited by Mortimer J. Adler, translated by G. R. G. Mure, seventh printing. Chicago: Encyclopædia Britannica.

Aristotle. 2003b. *History of Animals*. First published circa 350 BCE. Great Books of the Western World, Volume 8. Edited by Mortimer J. Adler, translated by D'Arcy Wentworth Thompson, seventh printing. Chicago: Encyclopædia Britannica.

Aristotle. 2003c. *On Interpretation*. First published circa 350 BCE. Great Books of the Western World, Volume 7. Edited by Mortimer J. Adler, translated by E. M. Edghill, seventh printing. Chicago: Encyclopædia Britannica.

Armstrong, Herbert W. 1934. What is Going to Happen! *The Plain Truth*. Volume 1, number 5, June-July 1934.

Ashton, John F., editor. 2009. *In Six Days: Why Fifty Scientists Choose to Believe in Creation*. Green Forest: Master Books.

Augustine of Hippo. 2003. *The Confessions*. First published circa 400 CE. Great Books of the Western World, Volume 16. Edited by Mortimer J.

Adler, translated by R. S. Pine-Coffin, seventh printing. Chicago: Encyclopædia Britannica.

Barisic, Sonja. 1999. Antichrist Is Alive And A Jew, Falwell Says. *The Seattle Times.* 16 January 1999.

Benson, Herbert. 2005. Study of the Therapeutic Effects of Intercessory Prayer (STEP) in Cardiac Bypass Patients: A Multicenter Randomized Trial of Uncertainty and Certainty of Receiving Intercessory Prayer. *American Heart Journal*. Volume 151, number 4, May 2005.

Blackstone, W. E. 1916. The Times of the Gentiles. *The Weekly Evangel*. Number 139, 13 May 1916. St. Louis: Gospel Publishing House.

Boyer, Pascal. 2001. *Religion Explained: The Evolutionary Origins of Religious Thought*. New York: Basic.

Bragg, Melvyn. *In Our Time*. British Broadcasting Corporation (BBC). [audio podcast]

Calvin, John. 1845. *The Institutes of the Christian Religion, Volume 2*. First published 1536. Translated by Henry Beveridge. Edinburgh: Calvin Translation Society.

Cambridge University. 1922. *The Historical Register of the University of Cambridge*. Supplement 1911-20. Cambridge: Cambridge University Press.

Carroll, Rory. 2000. Pope Says Sorry for Sins of Church. *The Guardian*. 13 March 2000.

Cassidy, Edward Idris. 1998. *Commission for Religious Relations with the Jews: We Remember - A Reflection on the Shoah*. Papal Encyclicals Online. 16 March 1998.

Cran, William, and Ben Loetermann. 1998. Apocalypse. *Frontline*. Episode 1805W, 22 November 1998. Aired by Public Broadcasting System (PBS). [audio transcript]

Dalrymple, G. Brent. 1991. *The Age of the Earth*. Stanford: Stanford University Press.

Darwin, Charles. 2003a. *The Descent of Man, and Selection in Relation to Sex*. First published 1871. Great Books of the Western World, Volume 49. Edited by Mortimer J. Adler, seventh printing. Chicago: Encyclopædia Britannica.

Darwin, Charles. 2003b. *The Origin of Species by Means of Natural Selection*. First published 1859. Great Books of the Western World, Volume 49. Edited by Mortimer J. Adler, seventh printing. Chicago: Encyclopædia Britannica.

Dawkins, Richard. 2009. *The Greatest Show on Earth: The Evidence for Evolution*. New York: Free Press.

Dawkins, Richard. 2018. *Science and the God Delusion: A Conversation with Richard Dawkins*. UHI Millennium Institute. 2 April 2008.

Dennett, Daniel. 1995. *Darwin's Dangerous Idea: Evolution and the Meanings of Life*. New York: Simon & Schuster.

Descartes, René. 2003. *Discourse on the Method of Rightly Conducting the Reason.* First published 1637. Great Books of the Western World, Volume 28. Edited by Mortimer J. Adler, translated by Elizabeth S. Haldane and G. R. T. Ross, seventh printing. Chicago: Encyclopædia Britannica.

Epp, Claas. 2018. *Global Anabaptist Mennonite Encyclopedia Online*.

Festinger, Leon, Henry W. Riecken, and Stanley Schachter. 2014. *When Prophecy Fails: A Social and Psychological Study of a Modern Group that Predicted the Destruction of the World*. Radford: Wilder Publications.

Feynman, Richard. 1994. *The Character of Physical Law*. First published 1965. New York: Modern Library.

Ford, Derek C., and Carol A. Hill. 1999. Dating of Speleothems in Kartchner Caverns, Arizona. *Journal of Cave and Karst Studies*. Volume 61, number 2, August 1999, pp. 84-88.

Gallagher, Delia, and Daniel Burke. 2016. *Pope Says Christians Should Apologize to Gay People*. Cable News Network (CNN). 27 June 2016.

The Geological Society. [www.geolsoc.org.uk]

Gregory, T. Ryan. 2007. Orgel's Second Rule and Unbeatable Predation Tactics. *Genomicron*. Attributed to Leslie Orgel by Francis Crick, 13 November 2007.

Halley, Edmund. 1714. A Short Account of the Cause of the Saltness [sic] of the Ocean, and of the Several Lakes That Emit No Rivers; With a

Proposal, by Help Thereof, to Discover the Age of the World. *Philosophical Transactions*. Volume 29, number 244, pp. 296-300. London: The Royal Society.

Harris, Sam. 2011. *The God Debate II*. Public debate held at the University of Notre Dame, Illinois, 7 April 2011. [video]

Henke, Kevin R. 2005. Dr. Humphrey's Young-Earth Helium Diffusion Dates. *The TalkOrigins Archive*. 17 March 2005.

Horkheimer, Max. 1947. *Eclipse of Reason*. New York: Oxford University Press.

Hughes, Liz Rank. 1992. *Reviews of Creationist Books*. Second edition. Berkeley, The National Centre for Science Education.

Hume, David. 1960. *A Treatise of Human Nature*. First published 1740. Oxford: Clarendon Press.

Hume, David. 2003. *An Enquiry Concerning Human Understanding*. First published 1748. Great Books of the Western World, Volume 33. Edited by Mortimer J. Adler, seventh printing. Chicago: Encyclopædia Britannica.

Jantos, Marek, and Hosen Kiat. 2007. Prayer as Medicine: How Much Have We Learned? *The Medical Journal of Australia*. Volume 186, number 10, 21 May 2007.

Jefferson, Thomas. 1820. *Thomas Jefferson to José Correa da Serra, 10 April 1820*. The Thomas Jefferson Papers. Series 1: General Correspondence 1651-1827. Washington, D.C.: Library of Congress.

King James Bible Online. First published 1611. [kingjamesbibleonline.org]

Krauss, Lawrence M. 2015. Ben Carson's Scientific Ignorance. *The New Yorker*. 28 September 2015.

Lavoisier, Antoine. 1793. *Traité Élémentaire de Chimie, Volume 1*. Second edition. Paris: Cuchet Librarie.

Lazarus, William P., and Mark Sullivan. 2008. *Comparative Religions for Dummies*. Hoboken: Wiley.

Lenski, Richard. 2018. *Experimental Evolution*. Michigan State University. [myxo.css.msu.edu/ecoli/index.html]

Lents, Nathan H. 2019. Behe's Last Stand: The Lion of Intelligent Design Roars Again. *Skeptic*. Volume 24, number 1.

The Linnean Society of London. [linnean.org]

Liu, Tanzhou, and Wallace S. Broecker. 2000. How Fast Does Rock Varnish Grow? *Geology*. Volume 11, number 2, February 2000, pp. 183-186.

Lunardini, Virgil J. 1995. *Permafrost Formation Time*. Cold Regions Research and Engineering Laboratory (CRREL). US Army Corps of Engineers. Report 95-08, April 1995.

Luther, Martin. 1914. *Text der Predigten des Jahres 1546*. First published 1549. D. Martin Luthers Werke: Kritischer Gesamtausgabe, Volume 51. Sermon number 3, 17 January 1546. Weimar: Herman Böhlaus Nachfolger.

Manilius, Marcus. 1920. *Astronomicon, Book Four*. First published circa 10 CE. Commentary by A. E. Houseman. London: Grant Richards.

Mendez, Fernando L. 2013. An African American Paternal Lineage Adds an Extremely Ancient Root to the Human Y Chromosome Phylogenetic Tree. *The American Journal of Human Genetics*. Volume 92, 7 March 2013, pp. 454-459.

Morris, Henry M., editor. 1974. *Scientific Creationism*. San Diego: Creation Life Publishers.

Nichol, Francis D. 1945. *The Midnight Cry: A Defense of the Character and Conduct of William Miller and Millerites, Who Mistakenly Believed that the Second Coming of Christ Would Take Place in the Year 1844*. Washington, D.C.: Review and Herald Publishing.

Nye, Bill, and Ken Ham. 2014. *Is Creation A Viable Model of Origins?* Public debate held at Creation Museum, Kentucky, 4 February 2014. [video]

Petto, Andrew J., and Laurie R. Godfrey, editors. 2007. *Scientists Confront Creationism: Intelligent Design and Beyond*. New York: W. W. Norton.

Pope Apologises for Church Sins. 2000. British Broadcasting Corporation (BBC). 12 March 2000.

Pope Benedict XVI. 2009. *Caritas in Veritate*. Papal Encyclicals Online. 29 June 2009.

Pope Francis Condemns Child Sex Abuse and Church Cover-Ups. 2018.

British Broadcasting Corporation (BBC). 20 August 2018.

Pope Gregory I. 1911. *The Dialogues of Saint Gregory the Great*. First published circa 590 CE. Edited by Edmund G. Gardner, book 4, chapter 39. London: Phillip Lee Warner.

Pope John Paul II. 1992. *Ansprache von Johannes Paul II an die Teilnehmer der Vollversammlung der Päpstlichen Akademie der Wissenschaften*. Papal Encyclicals Online. 31 October 1992.

Pope John Paul II. 1995. *Letter to Women*. Papal Encyclicals Online. 29 June 1995.

Pope John Paul II. 1996. *Mensajes del Santo Padre Juan Pablo II a los Miembros de la Academia Pontificia de Ciencias*. Papal Encyclicals Online. 22 October 1996.

Pope John Paul II. 2000. *Day of Pardon*. Papal Encyclicals Online. 12 March 2000.

Pope Pius IX. 1864. *Syllabus of Errors*. Papal Encyclicals Online. 8 December 1864.

Pope Pius IX. 1870. *Decrees of the First Vatican Council*. Papal Encyclicals Online. Session 4, 18 July 1870.

Pope Sends First Email Apology. 2001. British Broadcasting Corporation (BBC). 23 November 2001.

Popper, Karl. 1965. *Conjectures and Refutations: The Growth of Scientific Knowledge*. First published 1963. Second edition. New York: Basic.

Ptolemy. 2003. *The Amalgest*. First published circa 150 CE. Great Books of the Western World, Volume 15. Edited by Mortimer J. Adler, translated by R. Catesby Taliaferro, seventh printing. Chicago: Encyclopædia Britannica.

Randi, James. 1990. *The Mask of Nostradamus*. New York, Charles Scribner's Sons.

Reiss, Steven. 2015. *The 16 Strivings for God: The New Psychology of Religious Experiences*. Macon: Mercer University Press.

Robertson, Pat. 1990. *The New Millennium: 10 Trends That Will Impact You and Your Family By The Year 2000*. Dallas: Word Publishing.

The Royal Society. [royalsociety.org]

Russell, Bertrand. 1930. *Marriage and Morals*. London: George Allen & Unwin.

Russell, Bertrand. 1950. *An Outline of Intellectual Rubbish*. Unpopular Essays. First published 1921. New York: Simon & Schuster.

Russell, Bertrand. 1997. *Last Philosophical Testament 1943-68*. The Collected Papers of Bertrand Russell, Volume 11. Edited by John G. Slater. London, Routledge.

Sagan, Carl. 1979. *Broca's Brain: Reflections on the Romance of Science*. New York: Ballantine.

Sagan, Carl. 1996. *The Demon-Haunted World: Science as a Candle in the Dark*. New York: Ballantine.

Sagan, Carl. 2000. Encyclopedia Galactica. *Cosmos: A Personal Voyage*. Episode 12. Aired by Public Broadcasting Service (PBS). [video]

Shakespeare, William. 2003. *A Comedy of Errors*. First published 1623. Great Books of the Western World, Volume 24. Edited by Mortimer J. Adler, seventh printing. Chicago: Encyclopædia Britannica.

Shoemaker, Eugene M. 1983. Asteroid and Comet Bombardment of the Earth. *Annual Review of Earth and Planetary Sciences*. Volume 11, May 1983, pp. 461-494.

Smith, Chuck. 1978. *End Times: A Report on Future Survival*. Paducah: Maranatha House.

Suyu, S. H. 2010. Dissecting the Gravitational Lens B1608+656. *The Astrophysical Journal*. Volume 711, 1 March 2010, pp. 201-221.

Teehan, John. 2010. *In the Name of God: The Evolutionary Origins of Religious Ethics and Violence*. Chichester: John Wiley & Sons.

Tegmark, Max. 2018. What Is and What Matters: A Conversation with Rebecca Goldstein and Max Tegmark. *Waking Up Podcast with Sam Harris*. Episode 120, 19 March 2018.

Tobin, Harold J. 2012. Tides in Theory and Practice. *Oceanography: Exploring Earth's Final Wilderness*. The Great Courses, course 1730, lecture 16. Chantilly: The Teaching Company. [video]

Tyson, Neil deGrasse. 2019. *Letters From an Astrophysicist*. New York: W. W. Norton.

Veenhoven, Ruut. 2007. *Measures of Gross National Happiness*. Presented at OECD World Forum, Istanbul, 30 June 2007.

Voltaire. 1900. *Candide ou l'Optimisme*. First published 1759. Paris: Jules Tallandier.

Walmesley, Charles. 1834. *The General History of the Christian Church, from Her Birth to Her Final Triumphant State in Heaven: Chiefly Deduced from the Apocalypse of St. John, the Apostle and Evangelist*. Published under the pseudonym Signor Pastorini. New York: John Doyle.

Washam, Erik. 2010. Lunar Bat-men, the Planet Vulcan and Martian Canals. *Smithsonian Magazine*. December 2010.

Whisenant, Edgar. 1988. *88 Reasons Why The Rapture Will Be In 1988.* Nashville: World Bible Society.

Williams, George E. 2000. Geological Constraints on the Precambrian History of Earth's Rotation and the Moon's Orbit. *Reviews of Geophysics*. Volume 38, 1 February 2000, pp. 37-59.

Willoughby, Janna R. 2008. Rapid Genetic Adaptation to a Novel Environment Despite Genome-Wide Reduction in Genetic Diversity. *Molecular Ecology*. 26 May 2008. Wiley Online Library.

Wilson, Ted N. C. 2014. God's Authoritative Voice. *Adventist Review*. 18 August 2014.

De omnibus dubitandum est

ORBIS
UNUM
PRESS

An imprint of La Jolie Vie, LLC

Made in the USA
Monee, IL
08 November 2026

67972902R00090